Sociology of the Visual Sphere

This collection of original articles deals with two intertwined general questions: what is the visual sphere, and what are the means by which we can study it sociologically? These questions serve as the logic for dividing the book into two sections, the first ("Visualizing the Social, Sociologizing the Visual") focuses on the meanings of the visual sphere, and the second ("New Methodologies for Sociological Investigations of the Visual") explores various sociological research methods to getting a better understanding of the visual sphere. We approach the visual sphere sociologically because we regard it as one of the layers of the social world. It is where humans produce, use, and engage with the visual in their creation and interpretation of meanings. Under the two large inquiries into the "what" and the "how" of the sociology of the visual sphere, a subset of more focused questions is being posed: what social processes and hierarchies make up the visual sphere? How various domains of visual politics and visuality are being related (or being presented as such)? What are the relations between sites and sights in the visual research? What techniques help visual researcher to increase sensorial awareness of the research site? How do imaginaries of competing political agents interact in different global contexts and create unique, locally-specific visual spheres? What constitutes competing interpretations of visual signs? The dwelling on these questions brings here eleven scholars from eight countries to share their research experience from variety of contexts and sites, utilizing a range of sociological theories, from semiotics to post-structuralism.

Dennis Zuev is a Research Fellow in the Centre for Research and Studies in Sociology (CIES-ISCTE-IUL) in Lisbon, Portugal.

Regev Nathansohn is the president (2010–2014) of the Visual Sociology Thematic Group working under the International Sociological Association, and a Ph.D. candidate in the Department of Anthropology at the University of Michigan, Ann-Arbor.

Routledge Advances in Sociology

For a complete list of titles in this series, please visit www.routledge.com

61 **Social Theory in Contemporary Asia**
Ann Brooks

62 **Foundations of Critical Media and Information Studies**
Christian Fuchs

63 **A Companion to Life Course Studies**
The Social and Historical Context of the British Birth Cohort Studies
Michael Wadsworth and John Bynner

64 **Understanding Russianness**
Risto Alapuro, Arto Mustajoki and Pekka Pesonen

65 **Understanding Religious Ritual**
Theoretical Approaches and Innovations
John Hoffmann

66 **Online Gaming in Context**
The Social and Cultural Significance of Online Games
Garry Crawford, Victoria K. Gosling and Ben Light

67 **Contested Citizenship in East Asia**
Developmental Politics, National unity, and Globalization
Kyung-Sup Chang and Bryan S. Turner

68 **Agency without Actors?**
New Approaches to Collective Action
Edited by Jan-Hendrik Passoth, Birgit Peuker and Michael Schillmeier

69 **The Neighborhood in the Internet**
Design Research Projects in Community Informatics
John M. Carroll

70 **Managing Overflow in Affluent Societies**
Edited by Barbara Czarniawska and Orvar Löfgren

71 **Refugee Women**
Beyond Gender versus Culture
Leah Bassel

72 **Socioeconomic Outcomes of the Global Financial Crisis**
Theoretical Discussion and Empirical Case Studies
Edited by Ulrike Schuerkens

73 **Migration in the 21st Century**
Political Economy and Ethnography
Edited by Pauline Gardiner Barber and Winnie Lem

74 **Ulrich Beck**
An Introduction to the Theory of Second Modernity and the Risk Society
Mads P. Sørensen and Allan Christiansen

75 **The International Recording Industries**
Edited by Lee Marshall

76 **Ethnographic Research in the Construction Industry**
Edited by Sarah Pink, Dylan Tutt and Andrew Dainty

77 **Routledge Companion to Contemporary Japanese Social Theory**
From Individualization to Globalization in Japan Today
Edited by Anthony Elliott, Masataka Katagiri and Atsushi Sawai

78 **Immigrant Adaptation in Multi-Ethnic Societies**
Canada, Taiwan, and the United States
Edited by Eric Fong, Lan-Hung Nora Chiang and Nancy Denton

79 **Cultural Capital, Identity, and Social Mobility**
The Life Course of Working-Class University Graduates
Mick Matthys

80 **Speaking for Animals**
Animal Autobiographical Writing
Edited by Margo DeMello

81 **Healthy Aging in Sociocultural Context**
Edited by Andrew E. Scharlach and Kazumi Hoshino

82 **Touring Poverty**
Bianca Freire-Medeiros

83 **Life Course Perspectives on Military Service**
Edited by Janet M. Wilmoth and Andrew S. London

84 **Innovation in Socio-Cultural Context**
Edited by Frane Adam and Hans Westlund

85 **Youth, Arts and Education**
Reassembling Subjectivity through Affect
Anna Hickey-Moody

86 **The Capitalist Personality**
Face-to-Face Sociality and Economic Change in the Post-Communist World
Christopher S. Swader

87 **The Culture of Enterprise in Neoliberalism**
Specters of Entrepreneurship
Tomas Marttila

88 **Islamophobia in the West**
Measuring and Explaining Individual Attitudes
Marc Helbling

89 **The Challenges of Being a Rural Gay Man**
Coping with Stigma
Deborah Bray Preston and Anthony R. D'Augelli

90 **Global Justice Activism and Policy Reform in Europe**
Understanding When Change Happens
Edited by Peter Utting, Mario Pianta and Anne Ellersiek

91 **Sociology of the Visual Sphere**
Edited by Regev Nathansohn and Dennis Zuev

Sociology of the Visual Sphere

Edited by Regev Nathansohn
and Dennis Zuev

NEW YORK   LONDON

First published 2013
by Routledge
711 Third Avenue, New York, NY 10017

Simultaneously published in the UK
by Routledge
2 Park Square, Milton Park, Abingdon, Oxfordshire OX14 4RN

First issued in paperback 2014

Routledge is an imprint of the Taylor and Francis Group, an informa business

© 2013 Taylor & Francis

The right of Regev Nathansohn and Dennis Zuev to be identified as the authors of the editorial material, and of the authors for their individual chapters, has been asserted in accordance with sections 77 and 78 of the Copyright, Designs and Patents Act 1988.

All rights reserved. No part of this book may be reprinted or reproduced or utilised in any form or by any electronic, mechanical, or other means, now known or hereafter invented, including photocopying and recording, or in any information storage or retrieval system, without permission in writing from the publishers.

Trademark Notice: Product or corporate names may be trademarks or registered trademarks, and are used only for identification and explanation without intent to infringe.

Library of Congress Cataloging-in-Publication Data
Sociology of the visual sphere / edited by Regev Nathansohn and Dennis Zuev.
   p. cm. — (Routledge advances in sociology ; 91)
Includes bibliographical references and index.
1. Visual sociology. I. Nathansohn, Regev. II. Zuev, Dennis.
HM500.S63 2013
301—dc23
2012032647

ISBN 978-0-415-80700-5 (hbk)
ISBN 978-1-138-92077-4 (pbk)
ISBN 978-0-203-06665-2 (ebk)

Typeset in Sabon
by IBT Global.

Contents

List of Figures ix
List of Tables and Graphs xi

1 Sociology of the Visual Sphere: Introduction 1
  REGEV NATHANSOHN AND DENNIS ZUEV

PART I
Visualizing the Social, Sociologizing the Visual

2 The Limits of the Visual in the "War without Witness" 13
  PAVITHRA TANTRIGODA

3 From a Slight Smile to Scathing Sarcasm: Shades of Humor in Israeli Photojournalism 25
  AYELET KOHN

4 Sociology of Iconoclasm: Distrust of Visuality in the Digital Age 42
  ŁUKASZ ROGOWSKI

5 Picturing "Gender": Iconic Figuration, Popularization, and the Contestation of a Key Discourse in the New Europe 57
  ANNA SCHOBER

PART II
New Methodologies for Sociological Investigations of the Visual

6 Production of Solidarities in YouTube: A Visual Study of Uyghur Nationalism 83
  MATTEO VERGANI AND DENNIS ZUEV

7	On the Visual Semiotics of Collective Identity in Urban Vernacular Spaces TIMOTHY SHORTELL AND JEROME KRASE	108
8	Representing Perception: Integrating Photo Elicitation and Mental Maps in the Study of Urban Landscape VALENTINA ANZOISE AND CRISTIANO MUTTI	129
9	Operations of Recognition: Seeing Urbanizing Landscapes with the Feet CHRISTIAN VON WISSEL	160
	Contributors	183
	Index	187

Figures

3.1	Elections, Tess Scheflan, 2009.	31
3.2	Tel-Aviv Beach, Alex Levac, 2005.	32
3.3	Ariel Sharon, Micha Kirshner, 1983.	33
3.4	Tufah village, Pavel Wolberg (printed in Wolberg, 2006).	37
5.1	Daniela Comani, from the series "EINE GLÜCKLICHE EHE" (A HAPPY MARRIAGE): #2, 2003, © Daniela Comani.	65
5.2	Jenny Saville, "Passage" (2005), © Jenny Saville.	67
5.3	"Jake the Rapper," Gender Bender Festival catalogue (2007), © Harald Popp and Jake the Rapper.	71
5.4	Graffiti, Berlin, Friedrichshain, 2009, © Henning Onken.	73
6.1	Language used in YouTube videos about Uyghurs.	89
6.2	Clash of Chinese and Uyghur symbols in nationalist ideological code.	93
6.3	Map symbolism in nationalist (pan-Turkist) ideological code.	94
6.4	The iconic clash of Turkic and Chinese civilizations.	95
6.5	Images of purity and nature in human rights ideological code.	97
6.6	Militants from Islamic Party of East Turkestan issue a warning to the Chinese authorities.	98
7.1	Phatic signs of collective identity.	117
7.2	Expressive signs of collective identity.	120
7.3	Conative and poetic signs of collective identity.	122
8.1	Mental map overlapping with the geographical map extracted from GoogleEarth.	139
8.2	The interface of the Flash application that generates hybridization between the mental map and the geographical map showing the correspondences between the elements present in the two maps.	140
8.3	Top: research area and interviews' localization; bottom: photo elicitation set.	142

8.4 Top: mental map with high correspondence. Bottom: mental map with north and south completely reversed made by interviewee GR3-Int05-B. 147
8.5 The two sets of photos (A and B) used for the photo elicitation. 150
9.1 The Pylon and the Bank Account Monuments. 173
9.2 The Manhole Monument and Monumental Line. 174
9.3 The Tyre and Pyramid Monuments. 175

Tables and Graphs

TABLES

8.1	Photos' Recognition	144
8.2	Photos' Positioning on the Mental Map	144

GRAPHS

8.1	Degree of correspondence and number of places marked on the mental maps.	148

1 Sociology of the Visual Sphere
Introduction

Regev Nathansohn and Dennis Zuev

This collection of original articles deals with two intertwined general questions: what is the visual sphere, and what are the means by which we can study it sociologically?[1] These questions serve as the logic for dividing the book into two sections: the first ("Visualizing the Social, Sociologizing the Visual") focuses on the meanings of the visual sphere, and the second ("New Methodologies for Sociological Investigations of the Visual") explores various sociological research methods to getting a better understanding of the visual sphere. We approach the visual sphere sociologically because we regard it as one of the layers of the social world. It is where humans produce, use, and engage with the visual in their creation and interpretation of meanings.[2] Under the two large inquiries into the "what" and the "how" of the sociology of the visual sphere, a subset of more focused questions is being posed: What social processes and hierarchies make up the visual sphere? How are various domains of visual politics and visuality being related (or being presented as such)? What are the relations between sites and sights in the visual research? What techniques help visual researcher to increase sensorial awareness of the research site? How do imaginaries of competing political agents interact in different global contexts and create unique, locally specific visual spheres? What constitutes competing interpretations of visual signs? The dwelling on these questions brings here eleven scholars from eight countries to share their research experience from variety of contexts and sites, utilizing a range of sociological theories, from semiotics to post-structuralism.

This book joins a growing literature on visual sociology and anthropology.[3] In this introduction, it is our intention to highlight what we see as the contribution of this book to the larger body of scholarly literature on the topic. Ultimately, it is our intention here to "normalize" visual sociology as an integral component of sociological study. What this collection proposes is that the visual is everywhere, no matter where you "look." Indeed, and much like every other field of sociological inquiry, there are unique characteristics also to the visuals. These are well expressed in every research context of the chapters in this book. However, these characteristics always—although to changing degrees—interact with other sociological factors such as class, gender, ethno-racial power relations, and institutionalization, to name a few. This is the reason we suggest to normalize the visual within the sociological research.

Suggesting to normalize the visual within sociology does not assume that sociologists have yet discovered the visual. In fact, visuality and the visual sense has since long time been discussed as sociologically relevant and significant, for example, to our understanding of mobilities (Urry 2000) and of social existence (Sztompka 2008). Moreover, the theory of iconic consciousness, proposed by Jeffrey Alexander (see, e.g., Alexander 2010, Alexander et al. 2012), intends to recover the invisible strands of meaning in the aesthetic of mundane materiality of everyday life arguing that everyday experience is iconic.

However, despite what is widely conceived as the totality of the visual in the modern world, the verbal register in mainstream sociology still maintains its ascendancy over the visual one. Nevertheless, at the same time we are exposed to new communities of visual practices, such as social networking sites and YouTube, which emerge alongside the more classic social practices of visual interaction (and visual practices of social interaction). This constantly shifting field of enquiry requires continuous reflexivity and an ongoing development of research and observation tools.

The authors in this book all problematize different aspects of the visual sphere, from production and circulation of images, to their various framings by different actors for their individual and political purposes. In each of the chapters of this book there is an interplay between the general and the specific, where various theoretical and methodological aspects of the visual analysis are being contextualized in individual case studies. The conjunction of all researches shared here teaches us on the necessity to conceive the visual sphere as a multitude of relations between the images, their agency, and politics, whereby meanings are created and negotiated. The image, we learn, has an agency which extricates it from the confining status of a mere representation to-be-interpreted. It is also an active force directed both at the audience and at the producer, but also at institutions which are involved in circulating it. Overall, the chapters in this book establish the notion that the visual sphere—where images are produced, circulated, interpreted, reproduced, and re-imagined—is an active social force in both regulating human relationships as well as in subverting it. In the analysis of the visual sphere it is therefore paramount to put to scrutiny the location of the visual in the interactions between human agency and institutional constraints. The authors in this collection examine such interactions in various empirical settings, from urban landscapes to collective identities, and from modes of production of images to the means by which they become visible through various media.

Our main argument in this book, therefore, is that in order to better understand the social world we cannot overlook the visual sphere. The first part of the book teaches us both on what we can find in the visual sphere,

Sociology of the Visual Sphere 3

as well as what social mechanisms are at work in creating, maintaining, and subverting it. The second part exposes us to methodologies of learning and analyzing the visual data we collect and produce. In both parts of the book, although the authors highlight the visual aspects of social life, they also remind us—whether implicitly or explicitly—that the visual cannot stand alone in the desire to understand the social. Therefore, we wish to emphasize here how in some observations shared in this book the visual signifies other social mechanisms, whereas in other cases the visual is regarded as a trigger for social transformations.

The first part of the book opens with Tantrigoda's research (Chapter 2) which brings to the fore a discussion on the truth value of visual representations. Building on the case of media representations of the conflict between the government of Sri Lanka and the Liberation Tigers of Tamil Eelam (LTTE), Tantrigoda highlights the overarching assumption among the various actors in this conflict that there is a reality out there, but that this reality is also easy to be manipulated once subjected to visual representation by political actors. In the examples analyzed by Tantrigoda we see how in some cases the politics of representation and the controversy over the reliability of images can overshadow the discussion on the represented conflict to the degree that it becomes an integral part of the conflict itself.

Much like other social spheres, the visual sphere, therefore, should be understood in context, and it is our responsibility as critical viewers to learn as much as we can about the political conditions which enable specific mechanisms of production and dissemination of images in order to develop a careful reading of the visual.

The notion of "manipulation" is problematized in the following chapter, where Kohn (Chapter 3) focuses on how manipulations are in fact embedded within the modes of production of visual representations. Particularly, she shows us how a photographer's gaze can frame a certain situation to be seen in the photograph as a humorous scene. Analyzing photographs taken by Israeli photojournalists, Kohn focuses on the gap between the "reality" and its representations by means of creative intervention of the photographer who signals incongruousness in the scene. The absurdities such photographs present may then allow the viewers to develop critical observation and to rethink scenes they may be familiar with. This is how the act of production of the visual—by means of looking, framing, and presenting—may change the ways we look at our surrounding, interpret it, and react to it politically. This, however, could be achieved only when the photographer and the viewers share similar social conventions. According to Kohn, the perceived synchronicity of everything that is captured within the photographic frame precludes the common mechanism in humor—that of gradually arriving to the punch. Thus, to be successful it must be compensated by building on previous and shared knowledge of the photographer with the viewers. In her examination of the various gazes juxtaposed in some of the images—the gaze of the photographer, the gaze of the photographed, and the gaze of the viewers—Kohn suggests, for example, that what she calls "the contemptuous gaze" can serve as another mechanism that not only compensates on the synchronic interpretation of the image but also opens the possibility for critical self-examination. This is the moment when an image may serve as a trigger for reexamination of other social elements, which—depending on other social factors—can lead to social transformations.

An examination of the relations between social transformations and the visual sphere is shared by Rogowski (Chapter 4), who analyzes contemporary formations of iconoclasm (*transformation iconoclasm*, *digital iconoclasm*, and *everyday iconoclasm*) and situates their practices within the history of distrust and destruction of visual images. Rogowski's analysis suggests considering iconoclasm as a prism through which we can examine the changes of functioning of the visual sphere, the changing ontological status of the visual, and the changing relations between sites, sight, and other senses. Rogowski's definition of certain contemporary digital manipulations of images as iconoclasm is a thought-provoking exercise which assumes that images still retain powers beyond their materiality and that destroying them can have an effect in the world. In a way, Rogowski's analysis paints the visual sphere as a battlefield where images are both the targets and the means of achieving the targets.

While Rogowski invites us to think about digital iconoclasm as a contemporary image-based mechanism for intervention in the social sphere, Schober (Chapter 5) offers an analysis which shares similar assumptions with regards to the power of images in the social sphere, but offers different practices for achieving it. In the examples she analyzes, it is not iconoclasm that the social actors (visual artists, in her case) choose as their course of action, but the creative forms of visual innovations. Schober is particularly interested in the ways whereby the concept of "gender" appears in the public sphere, in the struggles around it, and in the options the visual sphere can offer in generating a social change. Schober investigates a variety of popular and artistic adoptions of concepts of gender in diverse European cultural contexts, and focuses on the visually based attempts to rearrange the public discourse from focusing on "women" to the problematization of gender. Images of gender, Schober argues, perform a double-edged role: on the one hand, they increase the public profile of gender; on the other, these images show how sociological concepts and interpretations can be put to crisis. The cases examined by Schober may lead to distrust mainstream visualizations of gender where the concept is presented as a seamless and non-conflicting reality. Such iconic images of gender may clash with the empirical reality of multiple gender configurations.

Schober thus shows how the visual sphere can serve as a platform for testing, exposing, and playing with new—and old—social concepts. It allows for the imagination to be public, to be displayed, and to be discussed and negotiated by means of verbal and visual languages. The visual sphere can thus serve as a playground for ideas yet to be institutionalized. Once the image is being "freed" from the constraints of the social norms, she argues, what might seem familiar—such as concepts of masculinity, femininity, androgyny, sexuality, family, and genealogy—could be challenged and confronted with the much more chaotic experiences that the social order aspires to suppress.

What these approaches assume is that the visual is more than a medium. Indeed, it can serve as a medium for maintaining the normative social order, or for challenging it. The chapters in this section show that the visual is also an integral part of the wider social sphere, and has its own version of chaos and order. The visual is perceived as being more than a medium because it not only conveys messages; it also acts and is acted upon. But the visual is not alone, as the chapters in the second section of this collection show so well. While Schober discusses several visually-based attempts to change the way we think about "gender," such a focus on the visual (how gender looks) hinders experiences related to other sensual or to non-sensual aspects of gender, which are not necessarily visual (or not always visual, or not in causal relations with the visual). Turning from the sociology of the visual sphere to the sociology of "visualism," we may ask: if we assume that what is *material* is also visualizeable, then what is the visual status of that which is *ideal*? In the examples that Schober discusses, we *see* that what is imagined *can* be visualized, and thus materialized. Therefore, it is possible to argue that visualizing the imagined (or idealized) must be materially mediated. This argument, however, is being further problematized in each of the chapters in this collection, with every doubt they expose (e.g., in the debate over the veracity of images in Tantrigoda's chapter), and with every gap they point at (such as the temporal gap between a political revolution and the following iconoclastic events described in Rogowski's chapter, or the gap between people's perceptions of their environment and the maps they draw, as described in Anzoise and Mutti's chapter). Instead of offering a solution to the paradoxical position of the visual within the matrixes of the material/ideal and imagined/real, what the *sociology* of the visual sphere can offer is an elaboration on the human creative experiences in acting with and against the visual not only despite these paradoxes but also by utilizing them.

Some of the means by which such sociological investigation can be carried out are exemplified in the second part of the book, where the authors discuss contemporary research methods ranging from analysis based on photos taken by the researchers (and elaborating on how to take them, and how to store and organize them), through analyzing visuals (photographs, drawings, video clips) created by the research subjects, to analyzing the interaction with and interpretations of the visual world which surrounds us.

The second section opens with Vergani and Zuev's research (Chapter 6) which offers tools for visual analysis in the Web 2.0 age, thereby defining YouTube as a social space. They show how a study on Uyghur nationalism can benefit enormously from analyzing the various aspects of YouTube clips on the topic: the content and form of the clips; their modes and contexts of production, circulation, and reception; and the interactions between the clips' viewers and the network of YouTube users. Their chapter thus presents an innovative inquiry into current day's means by which political mobilization is taking place in a context of ethno-national conflict. Like other contributors to this book, Vergani and Zuev also show how the visual cannot stand alone. Thus, besides exploring the various ramifications of the topic under investigations (in their case, the technical means of producing YouTube videos and the competing ideological codes expressed in them, to name a few), they also show how these could be integrated with other elements related to the topic (such as the linguistic aspects of the YouTube videos). Additionally, and much like Anzoise and Mutti (Chapter 8), they also offer a mixed methods approach and demonstrate how quantitative and qualitative analyses feed one another and contribute to the fine tuning of the research protocol, as well as to the crystallization of the main sociological argument. It is this research method that leads them, for example, to uncover the optimism of Uyghur nationalists to engender change and internationalize the conflict between Han Chinese and Uyghurs via YouTube.

Shortell and Krase (Chapter 7) also utilize the Internet, although for a different purpose. Throughout the recent years they have created an online public visual database which contains thousands of images available for everyone to explore and analyze. This database contains images Shortell and Krase keep producing for their study of glocalization in urban settings. Based on these researcher-produced images, Shortell and Krase study the various visual expressions of glocalization. Traveling between several global cities throughout the world (sometimes visiting the same city more than once) they took thousands of pictures of urban neighborhoods that serve as their raw data to be analyzed. In their contribution for this book they share part of their larger project, and focus on images they took in seven global cities. To show how urban vernacular neighborhoods in these cities change as a result of globalization they extend Jakobson's semiotics to be useful for an analysis of visual signs in the urban setting and couple it with symbolic interactionism. Both of these methods are being used here under the framework of the grounded theory, which allows them to reveal the manner in which local groups and individuals assert agency in a multitude of ways and intervene in the vernacular space by performing visible signs of their collective identity. These signs, once documented by visual researchers, create what Shortell and Krase call "the photographic survey," which allows researchers to conduct various kinds of analyses to enrich our knowledge of the social sphere.

Much like Vergani and Zuev, Shortell and Krase suggest paying close attention to visual signs which they perceive as being at the heart of social interaction to the degree that human intersubjective experience is dependent upon them. Focusing on photographs as a source for rich visual data

Sociology of the Visual Sphere   7

(which remains in a fixed form) has in itself a lot of advantages. Nevertheless, the spatial-temporal boundaries of the photograph also limit the ability to learn more about what is beyond it, and what preceded it. For example, these limitations make it difficult to directly infer from the visuals an argument about causality, about agents' intentionality and motivation, and about collective identities other than pointing at their visual expressions and at the ways these expressions may vary in different times and places. Whether there is a fixed and stable relationship between an identity and its visual expression, or whether identities themselves transform over time, and how, or whether a certain identity is stable or fluid in the manner people attribute it to themselves—are questions the answers to which cannot be based on visual evidence alone. Rather, such visual analysis can certainly teach on changes in visual signs and their interconnectedness in time and space, thereby stimulate more nuanced sociological questions. Thus, Shortell and Krases's useful elaboration on their method could be regarded as a call for a utilization of mixed methods in the sociological analysis, the visual being one of them.

Besides Vergani and Zuev's example, another innovative example for mixed methods approach is shared by Anzoise and Mutti (Chapter 8). They adopt a more reflexive methodological account based on their experience in conducting research on people's perceptions of their environment. In their account they discuss what could be presented as a methodology of multilayering the visual. They describe in detail the gradual and interactive process of creating the corpus of visual data to be analyzed. This process is then followed by the actual analysis wherein additional visual data is being used as a point of reference that the previous data is to be compared against. According to their methodological approach, in order to get closer to people's perceptions (on their environment, as in their case studies, or on any other topic) it is imperative to use several methodologies, the juxtaposition of which may overcome some of the limitations of each of these methodologies when used alone. Thus, a combination of maps drawn by research subjects, with photo-elicitation interviews using both researcher-generated photographs and found images, as well as with the visual surrounding at the location where interviews are being held—all serve as equally significant data to be integrated into the analysis. By utilizing such a mixed methods approach, and once conducted in a reflexive manner which accounts for all participants' positions in the process of co-production of data, not only are the researchers able to achieve a more fine-tuned analysis, but they also gain the ability to dwell on cases which may have otherwise be marginalized or neglected. These, for example, are the cases of interviewees who do not submit themselves to the researchers' request to draw a map of their surroundings. Only by also conducting an interview with them could the researchers understand that this, in fact, is an act of refusal derived from a regime of representation which assumes that drawing something—making it visible—confirms its existence to the degree that it might even legitimize it. Not drawing it, therefore, is a subversive act not directed against the research procedure but against an element in the researched phenomenon. Anzoise and Mutti's mixed methods approach convincingly shows how such cases should receive no less of researchers' attention than any other case.

The last chapter of the book elaborates on a different visually based subversive practice. In his inspiring essay, von Wissel (Chapter 9) diverts from the other chapters of this section in the sense that he focuses on the researcher's own intervention in the visual sphere. In that regard, von Wissel's approach, particularly because it is being examined in an urban setting, is antithetic to Shortell and Krase's approach which is utterly non-interventionist. The authors of both chapters argue to be doing "grounded work". However, whereas Shortell and Krase adopt the inductive grounded theory approach in collecting their data, von Wissel claims to be adopting a grounded perspective which is also embodied, in the sense that "you see with your feet," to get a better feeling of your surroundings and the various means of interaction with it. Because von Wissel's approach is not limited to conventional ways of seeing, rather to imagination-based seeing whereby the unremarkable turns monumental, it could be argued that his approach is a deductive one: seeing the familiar in the unfamiliar. Von Wissel's thought-provoking suggestion is based on his utilization of Robert Smithson's artistic "operations of recognitions" in 1967's New Jersey. Von Wissel takes this method to his own personal experiential-poetic enquiry in the urbanizing landscape of the municipality of Tecámac, Mexico. His experience then leads him to an intriguing discussion on the relevance of this approach to sociologists who are interested in non-traditional tools for seeing the urban, and for accessing the means by which the urban is being seen.

The chapters in this section therefore demonstrate different sociological techniques of making sense of the urban life by visual means. For the various authors in this section, urban landscapes not only provide a site for developing practices of seeing; they also provide assemblage of sights, which triggers sharpened observations of competing relations between dwellers and systems of signification, ideological codes, and signs of collective identity. Shortell and Krase, for example, employ photographic survey to record visual information at a particular (urban) place and time. They promote an approach according to which the researchers travel to create a systematic visual record of the physical and social sides of streetscapes. Von Wissel, on the other hand, employs Smithson's practice of "seeing with the feet," and Anzoise and Mutti not only travel themselves through Milan's urban space but also make an attempt to analyze the relations between the parts of Milan and its dwellers of different social origin. As Anzoise and Mutti argue, the imaginaries of the urban dwellers represented by the maps they draw (or refused to draw) show that identity of a place is always contested and multiple. This relationality of the identity of a place is further emphasized by von Wissel in his exploration of a (sub)urban site located thousands of miles away from Milan.

Contrary to the other authors in this collection, von Wissel emphasizes not only the way we observe as researchers, but also the ways we intervene and immerse ourselves, with all our senses, in the subject of our inquiry, to the degree that we become active agents in creating meanings. If this edited volume is read linearly, from beginning to end, we hope that von Wissel's chapter, the closing chapter of the book, will open creative research horizons within the widening field of visual sociology.

NOTES

1. The articles in this volume were first presented in 2010, in a set of sessions under the title "The Sociology of the Visual Sphere," which were held in Goethenburg, at the XVII World Congress of the International Sociological Association (ISA), and organized by ISA's Visual Sociology Thematic Group. This group was formed in 2007, had its first set of sessions in ISA's first Forum of Sociology in Barcelona, and was officially approved by ISA in 2009. The Visual Sociology Thematic Group also organized a number of academic sessions in the 40th World Congress of the International Institute of Sociology (IIS) in New Delhi (2012), and in ISA's Second Forum of Sociology in Buenos Aires (2012). The group issues biannual newsletters and organizes visual sociology workshops, and the group's membership body consists of sociologists from all continents.
2. Much like the private and public spheres, the visual sphere is also in continuous and multifaceted interactions with other spheres where meanings are created and negotiated.
3. Recent contributions to this field of knowledge include Krase (2011), Margolis and Pauwels (2011), Milne, Mitchell, and de Lange (2012), and Pink (2012), to name a few.

BIBLIOGRAPHY

Alexander, Jeffrey. 2010. "Iconic consciousness: the material feeling of meaning." *Thesis Eleven* 103(1):10–25.

Alexander, Jeffrey, Dominik Bartmanski, and Bernhard Giesen. 2012. *Iconic power: materiality and meaning in social life*. New York: Palgrave MacMillan

Krase, Jerome. 2011. *Seeing cities change: local culture and class*. Burlington, VT: Ashgate.

Margolis, Eric and Luc Pauwels. 2011. *The SAGE handbook of visual research methods*. Los Angeles: Sage.

Milne, E.-J., Claudia Mitchell, and Naydene de Lange. 2012. *Handbook of participatory video*. Lanham, MD: AltaMira Press.

Pink, Sarah. 2012. *Advances in visual methodology*. Los Angeles: Sage.

Sztompka, Piotr. 2008. "The focus on everyday life: a new turn in sociology." *European Review* 16(1):23–27.

Urry, John. 2000. *Sociology beyond societies: mobilities for the twenty-first century*. London: Routledge.

Part I
Visualizing the Social, Sociologizing the Visual

2 The Limits of the Visual in the "War without Witness"[1]

Pavithra Tantrigoda

A hallmark of (post)modern warfare is its excessive mediatization. Noting the propensity in contemporary media culture to reduce even war and violence to a spectacle, Baudrillard (1995) has famously claimed that the Gulf War actually did not take place, but was a carefully "simulated" media event. The very proliferation of the images of war has come to reveal the limits of visual data from conflict scenarios, eroding the epistemological privilege assigned to the visual as a reflection of the truth/real. A controversy over a video footage[2] from the recently concluded conflict between the state military of Sri Lanka and the Liberation Tigers of Tamil Eelam (LTTE), fueling allegations of war crimes against both sides, attests to the power of the visual image, as well as its tenuousness in an era saturated with images. This chapter examines the visual (re)presentations of the conflict between the separatist Tamil Tigers and the government of Sri Lanka as a site that reveals the limits, cracks, and unreliability of visual data. However, it also contends that as the only remaining visual testimonies to the "War without Witness," these images of war stand as a "real" that point toward the irreducible reality of the conflict, providing evidence to a historical event that occurred in time and space.

Since the inception of the studies on the visual, the power of the visual image and its ability to provide a "transparent" reflection of the world has been underscored. The Western epistemological tradition has equated knowledge with representations, which are judged based on their mimetic adequacy in depicting the external reality (Rorty cited in Griffin 2004). "The presupposition that the visual is a terrain of perception and experience untainted by precepts, concepts, language, has most often been buttressed by the idea of mimesis, characterized variously as the 'reflectionist' or 'picture theory of language' or the 'essential copy'" (Evans and Hall 1999:11). The idea of the "pure" self-referential image "free of representation, reference, and narrative" has continued to dominate the theories on the visual (Evans and Hall 1999:12).

Although the privileging of the visual has led to the use of visual data such as documentaries and photographs as verifiable sources of knowledge/truth, current research has come to recognize its limits and, consequently, its status as a medium that provides an unmediated access to the truth has been interrogated (Banks 2001). Current theories on the visual argue for a more complex process of interpreting images, underscoring that "meaning is constituted not in the visual sign itself as a self-sufficient entity, nor exclusively in the sociological positions and identities of the audience, but in the articulation between viewer and viewed, between the power of the image to signify and the viewer's capacity to interpret meaning" (Evans and Hall 1999:4). This complexity of the visual event is manifested in the visual depictions of conflict scenarios.

The visual representations of war and violence, which have proliferated since the Second World War, have acquired a special place in the studies on the visual. Visual images of warfare are claimed to bridge the gap between "firsthand experience and secondary witnessing" (Kansteiner cited in Griffin 2004). According to Sontag (2003:21), "the understanding of war among people who have not experienced war is now chiefly a product of the impact of these images." Despite the interrogation of the visual at a theoretical level, the visual representations of war appeal to the power that is traditionally accorded to the visual as providing a forceful reflection of the "real." States and guerrilla groups have understood the power wielded by the visual and have come to rely on it more than ever before for ideological and political purposes. The task of surveilling and visually representing the combat zone has been facilitated by the availability of cutting-edge technology, such as infrared cameras, spy helicopters, and satellite images. The increasing tendency of disseminating the images of war for mass consumption is manifested in the move by the states to embed reporters within combat units—a strategic move adopted by the Pentagon in its military offensive against Iraq. However, as Campbell (2003b:103) contends,

> the extensive management of the media coverage of war—as a conjunction of official restrictions and self-imposed standards—has for the most part diminished the verisimilitude of the resulting images. Constrained by the confines of the Coalition Media Center, reporters seeking an overview were (in the words of Michael Wolff) in danger of becoming little more than a series of "Jayson Blairs," constructing colorful accounts of scenes they had never witnessed.

"SCOPIC REGIMES" IN THE "WAR WITHOUT WITNESS"

Since the inception of the conflict between the state military and the LTTE, the visual media played a significant role in the narratives of war constructed and disseminated by both parties. These visual narratives, which were constructed through video footage and the photographs from the conflict zone, attempted to enforce particular modes of seeing the conflict. Feldman (2000:49) employs the notion of "scopic regime" to characterize

The Limits of the Visual in the "War without Witness" 15

"the agendas and techniques of political visualization: the regimens that prescribe modes of seeing and visual objects, and which proscribe or render untenable other modes and objects of perception." A scopic regime is an "ensemble of practices and discourses that establish the truth claims, typicality, and credibility of visual acts and objects, and politically correct modes of seeing" (Feldman 2000:49). Both the government of Sri Lanka and the LTTE strived to create competing scopic regimes to bring the conflict into political visibility. The warring parties endeavored to control and prescribe the ways in which the images that emerged from the war zone should be "seen" and interpreted.

The state media and the local media agencies that were authorized by the government of Sri Lanka provided extensive coverage to the conflict during its last phase via embedded media units with the state military forces. These video units brought to the living rooms of Sri Lankan viewers selective strands of a visual narrative of war. The video footage from the combat zone primarily depicted the military victories and the LTTE defeats, positioning the state and the military in the role of "the saviors" of the nation. In addition, it was showcased as a "humanitarian mission" that aimed to protect the interests of the Tamil civilians, who were trapped in the LTTE controlled areas. However, as Jelin (2003:27) claims, the projection of a heroic image of a certain group involves the silencing of mistakes and errors that can tarnish that image. The images that piece together an official narrative/memory of the war, in fact, are carefully framed to convey an infallible narrative of victory by the state, masking and effacing the violence, destruction, and death in the conflict.

On the other hand, the state relied on the use of modern technology to accuse the LTTE of war crimes and using the civilians trapped in the war zone as human shields. The state military sent spy helicopters to the "No War Zone,"[3] where civilians remained entrapped, to capture the LTTE activity within the area. These visuals were broadcast live to the international community and to foreign diplomats in an effort to legitimize the military solution to the conflict and demonstrate the falsity of accusations of "war crimes" against the state. At the same time, the powerful aerial footage of the mass exodus of the Tamil population fleeing to the government-controlled territories was used by the state to impart the "humanitarian nature" of their war effort to the world. The images of the LTTE cadres, firing at the Tamil civilians who were attempting to flee their territory, were captured by spy helicopters of the state military. These images were then used by the state as a potent visual testimony of the war crimes committed by the LTTE against the very community that they were avowedly fighting for.

The state aimed to control the narrative of war and broadcasted images that seemingly captured the reality of the conflict. These visuals thus produced and solidified a particular reality that catered to the interests of the state. Visual images are performative and can constitute the reality that they purport to reflect. As Baudrillard (1995) has noted, images are capable of producing the reality of war. These images of the conflict were performative in that they constituted and reinforced a discursive narrative that legitimized a militaristic solution to the conflict. The Sri Lankan government strove to justify its military offensive by deploying a visual narrative that seemingly reflected the "terrorism" and "barbarity" of the LTTE, obfuscating a complex political reality and the humanitarian facet of the conflict.

Photographic images perform an important role in representing the reality of war. As Sontag (1990:16) contends, "photographs may be more memorable than moving images, because they are a neat slice of time, not a flow. Television is a stream of underselected images, each of which cancels its predecessor. Each still photograph is a privileged moment, turned into a slim object that one can keep and look at again." However, certain critics have maintained that the photograph's impression of reality is a "mere mechanical trick, an artificial and deliberately staged 'effect of the real'" and, in producing the illusion of immediacy, photographs conceal the fact that "the medium itself has fundamentally shaped the habits of looking we employ to establish an event's veracity" (Baer 2002:2). However, in spite of this "critical debunking of photography's claim to be the most accurate, and hence the most truthful, mode of representation . . . we continue to perceive photographs as records of what is" (Baer 2002:3). In the conflict between the government of Sri Lanka and the LTTE, the photographic images were mobilized for the construction of coercive narratives of war by both parties. While the effects that these photographs produced in the viewer are yet to be ascertained, clearly, both video footage and photographs were deployed by the warring parties to achieve a similar end, that of establishing and solidifying their own narratives of the event. However, during the last phase of the combat, photographs stood as a supplement to video footage, which became the privileged medium in the construction of narratives of war for both parties and, hence, the major source of controversy. Thus, while photographs can be more memorable than videos as Sontag claims, videos were regarded in this conflict as a more "authentic" medium that can capture the immediate reality of war.

Nevertheless, photographic images were deployed in the construction of a powerful nationalist narrative by the state. The Sri Lankan Defense Ministry website contains a folder of photographic images of the military operation against the LTTE that function as powerful mnemonic sites and anchoring points in the official narrative construction of the event. The billboards celebrating the victory of the government that dominated the cityscape of Colombo attempted to inscribe the official narrative of the event in the minds of the public and were aimed at societal militarization. These billboards contained the photographs of military heroes holding the national flag at the site of the conquest and, in certain billboards, the image of the president also appeared in their midst. As Griffin (2004:390) contends, photographs tend to symbolize generalities, providing transcending frames of cultural mythology or social narratives. Counter to continuing popular perceptions of photographic media, photographs do not simply reflect events occurring before the camera but are inextricably implicated in the constructive process of discourse formation and maintenance.

These images, with their claims to a real and authentic representation of the event, can be considered as powerful means of constructing and reinforcing the state narrative of the event and naturalizing and normalizing the militaristic solution to the conflict.

However, presenting a nationalist narrative through media such as the Internet and through billboards has its own pitfalls. While allowing the nationalist ideology and state propaganda to intersect, the billboard presents an ambiguous medium, signaling an extent of complicity with contemporary global, capitalist structures (that will attenuate the avowed authenticity of the "nationalist" narrative of the state). The agency and control of the visual narrative of the conflict is wrested away from the state, in its incorporation into spaces that are shifting, ambiguous, potentially anarchic, and available for mass consumption. Further, these spaces allow for the intervention of counter-narratives. The availability of conflicting images and narratives of war in websites and news media destabilizes the authority of the state representation of the event, enabling competing narratives to emerge and circulate simultaneously. The state's attempt at constructing and disseminating a hegemonic nationalist narrative through modern global media is thus destabilized in the very process of constituting and articulating it.

Whereas the government of Sri Lanka has a more recent history of employing the visuals of the conflict, the LTTE used videos and photographs as an important part of the military machinery and the propaganda campaign from the very inception of the conflict. In fact, the LTTE had a separate unit of cadres who went to the battlefront and recorded the combat. Their videos portrayed the heroism of the LTTE leader and the cadres, and the barbarity of the state military. These were used in the propaganda campaigns of the LTTE—for the purpose of recruitment of soldiers, to appeal for financial aid from the Tamil Diaspora and the international community, and to justify the violent means they adopted to their own community. However, during the last stages of the final battle, the LTTE used the visual media for a different purpose, namely, seeking the intervention of powerful Western nations by seemingly exposing the violations of international humanitarian law by the state military. The video footage that the LTTE claimed to depict the shelling of hospitals in the No War Zone by the state military attracted a great deal of international attention. The LTTE appealed to the power and authority traditionally accorded to visual images as mirroring the real to urge the international community to intervene on their behalf and call for a ceasefire by depicting a humanitarian crises, which was conveyed through images of death, destruction, and suffering of the Tamil civilians in the No War Zone. These coercive images impelled the Western nations, as well as organizations such as the United Nations and Human Rights Watch to exert pressure on the government to halt the military offensive against the LTTE. In order to attain a political goal, the LTTE thus relied on the authority invoked by the visual images of war, which are seemingly an unmediated reflection of the real truth of the conflict.

THE LIMITS OF THE VISUAL

The visuals on war, and war photography in particular, according to Sontag (2003), can be regarded as potent purveyors of truth about the conflict and this desire for truth or naturalness of images dominates the public consciousness in contemporary era. In Sontag's view (2003:52), the Vietnam War marks a watershed in the industry of war photography, for "the practice of inventing dramatic news pictures, staging them for the camera, seems on its way to becoming a lost art." However, the visuals of war that emerged from recent conflicts provide a counterpoint to Sontag's claims, revealing that "photographs and other representative visual media are all capable of distorting external reality and they always have been" (Mirzoeff 2005:69).

The images of war that emerged from the conflict between the state and the LTTE manifest a tendency of manipulating visual images for politico-ideological ends. The warring parties used the visual media to construct and reinforce their versions of truth about the conflict and present this truth to the world. However, the reliability of visual data that emerged from the combat zone was eroded by the attempts to contest its validity by the warring parties, as well as "neutral" observers, and this continues to occur even today. A point of departure for these debates has been the barring of independent media agencies and unauthorized personal from the battlefield by the government of Sri Lanka, giving rise to speculations about what actually took place in the No War Zone during the last few months of escalated fighting. The authenticity of the video footage that was released by the LTTE sources, recording the state military's shelling of the hospitals in the No War Zone, was contested by the state using computer analysts, who exposed the manipulated nature of these visuals. In the view of the state, the powerful and appealing images of suffering Tamil civilians that appear in this video are a mere performance, a simulation of a humanitarian crisis to win the sympathy of the international community and to attain the LTTE's political goals. While the United Nations attempted to supplement the claims of the LTTE by providing satellite images that indicated the use of heavy weapons in the No War Zone, the state persisted on denying these accusations.

The Limits of the Visual in the "War without Witness" 19

Another controversial footage, apparently revealing the extra-judicial killings by the state military appeared on Channel Four after the war concluded, fueling allegations of war crimes against the state. Although the state promptly denied these charges, disclosing the "manipulated" nature of the footage, the LTTE supporters were of the view that it is indeed genuine. This video footage became a site where an intense politico-ideological battle was waged, thereby eroding the epistemological privilege assigned to the visual as providing access to the real or truth of what happened. Revealing the tenuousness of the visual in an age of technological miracles, Campbell (2003a:61) notes that "with the increased capacity for pictorial manipulation arising from the use of digital cameras and computer imaging, public laments about the associated loss of authority and truth are common." In the face of the attempts to manipulate the visual, its truth value and power to convince us diminish and erode. These images thus became mere simulations or fabricated versions of the real that appealed to the authenticity and the power of the visual to convince us of its truth.

As a result of the controversy surrounding the images from the conflict between the state and the LTTE, these videos and photographs became a terrain where multiple and conflicting meanings were generated, shaped by the viewers' politico-ideological-geographic location in relation to the conflict. Baer (2002:12) uses the term "ungovernability" to refer to a photograph's "openness and interpretive instability." It is possible to extend Baer's definition of ungovernability to include video footage of war and violence that is characterized by such an "interpretive instability." The intense struggle over the authority of interpreting the visuals of war and the incommensurable narratives of the state, the LTTE, and the "independent" viewers gave rise to incertitude and ambiguity and revealed the discursive nature of truths that were premised on these images. Consequently, these images provided the grounds to interrogate the truth value of the visual, for visual images are regarded as providing an unmediated access to the real.

Although visual representations tend to function as an index of the real, effacing the process of construction at work, the controversial nature of the visual images of the war between the state and the LTTE has enabled the constructed and "interested" nature of these visuals to be deliberately foregrounded. Visual forms that employ tenets of realism (i.e., photography, documentaries, classic realist films) efface the process of production at work, passing off as reflections of a transparent and unmediated reality. Consequently, the politics of inclusion and exclusion, the point of view from which the events are rendered, the framing of the visual material, and the constraints of the medium become largely invisible.

However, when broadcasting the images of this particular conflict, the "independent" media agencies such as CNN and Al Jazeera cautioned the viewers that because these images are originating from either the LTTE or the government sources, they cannot guarantee the credibility of these images. In fact, the controversy over the truth value of these images became a significant part of the narrative of the conflict that was broadcasted by the international media, shaping the way in which they disseminated these images. Consequently, what these visuals purported to represent about the war became less important than the controversy itself. The international media undermined the authority of what these visuals represented, instead of underscoring the politics of representation. When telecasting the video footage of this particular conflict, Al Jazeera newsreaders emphasized that as no independent news reporter was allowed to enter the scene of conflict, they could not verify the authenticity of what was being shown. These media agencies thus attempted to differentiate their own news productions from the visual material that emanated from interested, "unreliable" sources, in their avowed aim of bringing the viewer an "unbiased" and "neutral" portrayal of the conflict. Stripped of authenticity and credibility, the sights of violence and terror emerging from the battlefield were thus reduced to a "media spectacle" in the mainstream international media amongst the proliferating mass of visual images that can be easily consumed and forgotten, testifying to the limits of the visual in an era saturated with images.

"NOTHING"—"WITNESSING" AND THE INTERNATIONAL MEDIA

> For the images to become a source of true information, they would have to be different from the war. They have become today as virtual as the war itself, and for this reason their specific violence adds to the specific violence of the war. In addition, due to their omnipresence, due to the prevailing rule of the world of making everything visible, the images, our present-day images, have become substantially pornographic. Spontaneously, they embrace the pornographic face of the war. (Baudrillard 2006:87)

The conflict between the state military of Sri Lanka and the LTTE is referred to as the "War without Witness" in the international media, underlining the importance of the visual and, especially, the mediatized forms of visual in contemporary global culture. This particular conflict was named thus, not because of the dearth of actual witnesses or visual recordings of the event, but because the international media, as well as certain local media agencies, were barred from the war zone by the government of Sri Lanka, citing security reasons. Further, the censorship exercised by the state allegedly prevented an unbiased view of the conflict. However, the controversy over the video footage from this conflict attracted much political visibility and broadcasting time. Despite the contentious nature of the visual data that emerged from the site of war, it became a prime news item in the agendas of international media channels such as CNN, BBC, and Al Jazeera. The primary concern of the international media was to represent the realities of war that has eluded their cameras and to bring to light the fate of the civilians who were trapped in the No War Zone and, later, in the refugee camps.

However, international media's pursuit of the real of this conflict is very much a discursive real that is informed by hegemonic representational practices about the Third World. The manner in which the visual images from Sri Lanka's conflict were broadcasted in the Western media, displaying naked images of death and destruction, and thus emphasizing the horror as the real, functions to reinforce a discursive framework that constitutes Third World countries as without democracy, where real horrors happen every day. Žižek (2002) has drawn attention to the distinct ways in which international media agencies represent crises in the First World and the Third World. He claims that while naked images of death are shown when representing the conflicts in the Third World, these media agencies exercised caution not to do the same when showing, for instance, the attack on the World Trade Center by Al Qaeda terrorists. Such representational practices of the international media can be regarded as a part of the process of "othering" the Third World. Further, the tendency to sensationalize and commodify the narratives of war, violence, and suffering from the Third World is linked with the issues of power. Hall (1973:57) has correctly pointed out that "representational legitimacy remains inextricably tied to power, even if the links are complicated by layers of social hierarchy and specific historical relationships." Notwithstanding the fact that these videos were already a source of controversy, the authority of the visuals emanating from the Sri Lankan military sources were easily undermined in international media, because these originated from a Third World, less powerful nation, fighting for a cause that was not advocated by the powerful Western nations.

Once the conflict has concluded and the restrictions on media have lifted, international media strived to fill the gap left by the "War without Witness" by pursuing the narratives of the victims/survivors of the conflict. "The Future Holds Nothing" is one such program broadcast on CNN, in which the journalist Sara Sidner travels to the war-torn areas of Sri Lanka in a quest of victims/survivors of conflict. The horrors of the war are imaged by the presence of maimed victims/survivors in front of the camera, who relate their stories of suffering and loss during the conflict in gruesome detail. Raveendran Jenatha, a twenty-one-year-old female, who was trapped in the middle of the combat zone, recounts to CNN, "We got into a bunker, and a shell fell inside the bunker where we were taking shelter. My cousin died. Me? I lost my two legs." Minutes later, she adds, "I don't have an eye either." In its quest for the witnesses/survivors of war and their narratives, CNN is attempting to delve into the real of the conflict, which, for CNN, is represented by the humanitarian crisis. However, as Beegan and Kraus argue, "while claiming to be presenting the reality of the war, human interest journalism contributes to an important obfuscation of its political reality for the domestic audience, a visual obfuscation with long historical precedents" (cited in Malik 2006:82). In their exclusive focus on the victims, their affective histories, and the moral/ethical facet, such representations render a one-dimensional view of the conflict, undermining and effacing a complex political reality. Emphasizing the importance of providing a relevant political context in presenting the images of war, Sontag (1990:17) contends that "without a politics, photographs of the slaughter-bench of history will most likely be experienced as, simply, unreal or as a demoralizing emotional blow." The image itself is not capable of engendering change and "a photograph that brings news of some unsuspected zone of misery cannot make a dent in public opinion unless there is an appropriate context of feeling and attitude" (Sontag 1990:17).

Further, the narratives of survivors/victims of war register an absence that defies the hegemonic representational practices and the appropriations of their victimhood by media. The meanings that we read into their testimonies are structured around an absence that is at the heart of what is being shown or narrated. Baer (2002:12) contends that the photographs are marked by a "structuring *absence* [that] defines ... [traumatic] experiences." For him, trauma signifies the ungovernable, which, like the past, is an unstable referent (Baer 2002:107). Thus, "[p]hotographs of trauma present to the viewer representations that call into question the habitual reliance on vision as the principle ground for cognition" (Baer 2002:181). While Baer refers exclusively to photographs, the video interviews that CNN conducted with the victims of war in Sri Lanka seem to register a similar effect. While those who were trapped in the No War Zone during the conflict attempted to reconstruct their experiences of war through narratives that invoked their trauma and suffering, these partially signified the reality of their experience to the viewer, for language seems to provide an inadequate vehicle to capture the full extent of their trauma. Further, the visible markers of suffering such as scars and their tears when they were referring to the lost relatives and friends in the battle pointed to the brute reality of war for these survivors. However, these visible markers can be regarded as inadequate texts that mark an absence or an absent presence, for their pain and trauma remain partially signifiable. The images that convey these partial narratives and absences point toward the limits of representational activity. Rather than assigning overarching meanings to the ungovernable that expresses trauma and loss incurred in the conflict, the limits of these visuals and the structuring absences have to be acknowledged in representing them.

Sontag concludes *Regarding the Pain of Others* (2003:115) with the following: "Let the atrocious images haunt us. Even if they are only tokens, and cannot possibly encompass most of the reality to which they refer, they still perform a vital function. The images say: This is what human beings are capable of doing—may volunteer to do, enthusiastically, self-righteously. Don't forget." Although the meanings that are generated by the visuals of the conflict are irreducible to a moral/ethical dimension, it constitutes a vital part of it. As the only remaining visual testimonies to the "War without Witness," these images of each other's barbarities stand as a real that is irreducible and construct a narrative of death, destruction, and suffering that can only be partially signified. Notwithstanding the undermining of the visual, these images point toward the irreducible reality of the conflict, providing evidence to a historical event that occurred in time and space. The meanings that are generated in this in-between, unstable space of negotiation of the visual with an "unverifiable real" and the voids, traces, and excesses that it leaves behind are crucial for the socio-visual researcher in his/her quest for the truth in the realm of the visual.

NOTES

1. The conflict between the government of Sri Lanka and the Liberation Tigers of Tamil Eelam (LTTE) was referred to as the "War without Witness" in the international media, as unauthorized media agencies were barred from entering the battlefield during the last few months of escalated fighting.
2. The LTTE sources released a video footage to the international media that claimed to depict the shelling of the hospitals in the "No War Zone" by the state military in 2009, fuelling allegations of war crimes against the government. This became the source of a huge controversy, because No War Zone was declared as an area where no heavy weaponry could be used, because Tamil civilians remained trapped there along with the LTTE carders during the last few months of the conflict. The government contested the veracity of the footage, claiming that the visuals have been manipulated by the LTTE. However, because no independent media was allowed into the area, it remained impossible to verify the authenticity of these visuals.
3. "No War Zone" designates an area less than ten square miles in the northern part of Sri Lanka, where nearly 100,000 Tamil civilians remained trapped during the last few months of the conflict between the LTTE and the state military. A large number of LTTE carders, including their leader, were also cornered in the No War Zone along with the civilians and the LTTE was accused of using civilians as human shields. Because of the presence of the civilians, the state declared that it would not use heavy weaponry or aerial bombardment within the No War Zone.

REFERENCES

Baer, Ulrich. 2002. *Spectral evidence: The photography of trauma*. Cambridge, MA: MIT Press.
Banks, Marcus. 2001. *Visual methods in social research*. London: Sage.
Baudrillard, Jean. 1995. *The Gulf War did not take place*. Translated by Paul Patton. Bloomington: Indiana University Press.
Baudrillard, Jean. 2006. "War Porn." *Journal of Visual Culture* 5 (1): 86–88.
Campbell, David. 2003a. "Cultural governance and pictorial resistance: reflections on the imaging of war." *Review of International Studies* 29:57–73.
Campbell, David. 2003b. "Representing contemporary war." *Ethics & International Affairs* 17(2):99–108.

Evans, Jessica and Stuart Hall, eds. 1999. *Visual culture: the reader.* London: Sage.
Feldman, Allen. 2000. "Violence and vision: the prosthetics and aesthetics of terror." Pp. 46–78 in *Violence and subjectivity*, edited by V. Das. Berkeley: University of California Press.
Griffin, Michael. 1999. "The Great War photographs: constructing myths of history and photojournalism," Pp. 122–157 in *Picturing the past: media, history, and photography*, edited by B. Brennen and H. Hardt. Urbana: University of Illinois Press.
Griffin, Michael. 2004. "Picturing America's 'War on Terrorism' in Afghanistan and Iraq: photographic motifs as news frames." *Journalism* 5(4):381–402.
Hall, Stuart. 1973. "The determination of news photographs." Pp. 176–190 in *The manufacture of news: social problems, deviance, and the news media*, edited by S. Cohen and J. Young. London: Constable.
Hallin, Daniel. 1986. *The uncensored war: the media and Vietnam.* New York: Oxford University Press.
Hardt, H. 1991. "Word and images in the age of technology." *Media Development* 38(4):3–5.
Jelin, Elizabeth. 2003. *State repression and the labors of memory.* Minneapolis: University of Minnesota Press.
Malik, S. 2006. "The war in Iraq and visual culture: an introduction." *Journal of Visual Culture* 5(1):81–83.
Mirzoeff, Nicholas. 2005. *Watching Babylon: the war in Iraq and global visual culture.* New York: Routledge.
Sontag, Susan. 1990. *On photography.* New York: Anchor Books.
Sontag, Susan. 2003. *Regarding the pain of others.* New York: Farrar, Straus, and Giroux.
Zizek, Slavoj. 2002. "Welcome to the desert of the real." *The South Atlantic Quarterly* 101(2):385–389.

3 From a Slight Smile to Scathing Sarcasm

Shades of Humor in Israeli Photojournalism

Ayelet Kohn

INTRODUCTION

This paper explores various expressions and shades of humor as means of conveying social and political criticism in press photos, and examines this phenomenon as it is manifested in the printed and online Israeli press. An investigation of a representative sample of photographs published in the Israeli press since 2000 gives rise to two interlinked issues. One focuses on the attributes of photography as a medium, and with the challenge it presents to photographers when they come to use humor as a tool of conveying criticism. The second issue deals with the methods in which the Israeli reality, as well as the nature of the objects of representation and the intended audience of the photos, contributes to the epistemological questions raised by photography. It is an accepted notion that the medium influences the type of humor it can express, and that the humorous message changes according to the mode of its expression and the nature of the intended audience (Palmer 1987; Nielsen 1986; Shifman 2007). Therefore, the ways in which the humorous message of each image is encoded, and the range of possible interpretations by the viewer, must be investigated using epistemological, social, and cultural tools.

RESEARCH GOALS AND METHODS

This chapter presents a number of possible models for conveying humor in photojournalism, for charging them with a humorous tone, or for inviting the viewer to interpret them as humoristic comments expressing social or political criticism. Viewed from this perspective, press photos can shed light on the present role of visual images in the media as means of shaping consciousness, as designators of differential knowledge and power wielded by various actors, and as texts reflecting streams and approaches in contemporary photography (Elkins 2002).

I will begin with a preliminary discussion of the dilemmas presented by photography as a medium for expressing humor, and of the challenges it poses in understanding the creation and interpretation of the photos.

Various aspects of humor, evident mainly in ironical and critical messages, will form the basis for a systematic analysis, both semiotic and verbal, of photos employing humor, focusing on different levels of discrepancy and conflicting meanings. Some aspects of Sperber and Wilson's (1989, 1991) theory of echoic mention will be applied in the discussion of visual images and the dividing of meaning within their space.

Most of the photographs examined in this study could be attributed to the "decisive moment" genre in photography, whose characteristics in terms of narrative, form, and content allow a fruitful discussion of humor as a tool of conveying criticism. Because a humorous expression is rather elusive and challenging to define, the second part of the chapter will focus on two types of photographs: ones that incorporate written text, and ones presenting a situation that can be identified as humorous, and especially ironic (Scott 2004). In both types, it is the interplay between the photo's form and content, in all its various aspects, that triggers and shapes the critical message (Jay 2002).

Finally, I will suggest that a central concept in photography—the gaze—has a major role in signaling and organizing situations that are perceived as implausible and presented from a critical perspective. Understanding the relations between the photographer's hand and the gaze, and the perception of the photographer as a hunter (Berger 2003) but also as one who is himself or herself an object of observation, will illuminate the complexities involved in using shades of humor (and especially in signaling the incongruous) in political photojournalism.

As part of my analysis, I will discuss the characteristics of the humor on a spectrum ranging from "a slight grin" to "bitter contempt." I will focus on the methods in which shades of humor designate situations as absurd, and direct the viewer toward critical self-examination. To this end, I will perform a visual and content analysis of a sample of photographs published since 2000 in the Israeli press and in collections of press photos compiled into albums. The research is based on social approaches to semiotic analysis of visual images (Jewitt and Oyama 2001; Lister and Wells 2001). It should be noted that some of the photos have been displayed in exhibitions and have thus acquired a complex status: disassociated from the immediate context of documenting a concrete event, they have become marketable works of art. This affects the degree to which they can be perceived as subversive and critical, two central characteristics of humor (Douglas 1975). Most of the works to be discussed are by a number of well-known Israeli photojournalists who have received recognition by the establishment, such as Pavel Wolberg, Alex Levac, and Micha Kirshner. The recognition they have acquired contributes to their status as designers and leaders of public opinion, but also influences the character of their works, which are aimed (even if unconsciously) at resonating with a broad consensus and gaining public acclaim. The works of these photographers observe the political and social reality of Israel in the twenty-first century.

HUMOR IN PHOTOGRAPHY AND A COMMENT ON HUMOR AND POLITICS IN ISRAEL

The main problem in defining the characteristics and uses of humorous tones in photography—whether comical, amusing, intriguing, funny, or even shockingly absurd—is anchored in two frames of reference: epistemological and contextual. The synchronic character of photography inherently limits the possibilities unfolding a narrative or delineating a process. Both the artist and the viewer have difficulty in creating, identifying, and deciphering narrative components that are familiar from humorous texts, such as the sudden plot twist, the surprising punch line, or the absurd situation whose exaggerated aspects are clarified gradually as the story unfolds. Therefore, comical aspects or components in a photo are based on social conventions valid at a certain time and place. They rely on the assumption that most of the discourse-participants will identify some feature or combination in the photo as "funny," "comical," or as evoking a sense of contempt intended as criticism.

A search for the string "funny photographs" on the Internet will yield a range of images and slideshows that can be categorized in several different ways, reflecting diverse perceptions of humor (on the semiotic classification of humor see Palmer 1987:75–80, and on shades of irony, see Muecke 1973). One category is photos showing figures and situations that are immediately recognizable as incongruous or exaggerated in their way of presentation. These photographs involve distortion, or the exaggeration and highlighting of some feature or flaw in the outward appearance of the subject(s) represented. They are inspired by and close in genre to caricatures and grotesques, and they sometimes (although not always) share the critical function of caricatures. Another category is of photos that tell a "visual joke" or show a situation that violates behavioral norms. These images also echo familiar comical genres, like those involving hybrids (e.g., animals dressed as humans) and violation of behavioral norms (e.g., a photograph showing the "slapstick moment" of a pie about to be cast into someone's face, which triggers in the viewer a sense of relief based on a feeling of *schadenfreude*; see Zuromskis 2008). Yet another category is that of photographs evoking an affectionate (rather than derisive or ironic) smile, for example, those typically yielded by a search for "funny photographs with children" or "funny photographs with animals." It should be noted that all these comical photographs are meant to trigger an almost reflexive response that does not evoke further thought, like tickling. The boundaries between the categories are blurred, and we can identify many combinations of humor types that originally belonged to one specific genre, but currently contribute to creating a new understanding and definition of what is funny in a given set of circumstances.

As a rule, humorous genres encapsulate or highlight various aspects of the society they observe: prohibitions and fears, customs, and norms.

Example of this are presented in Kirshenblatt-Gimblett (1972), who discusses humorous jokes and stories of Jewish immigrants reflecting the difficulties of immigration, the struggle to preserve Jewish customs, and the timbre of Jewish community life.

An example for a photograph which enables the viewer to draw (certain) amusement from the conflicting readings, depending on the reader's familiarity with the Israeli reality and with the particular context, is one taken by Galy Tibbon during the 2008 Israeli parliamentary elections. The photograph shows an election sign which introduces Avigdor Liberman, chairman of the anti-orthodox Israel Beitenu (Hebrew: "Israel our home") party. The sign reads, "No Citizenship without Loyalty." A family of orthodox Jews passes under the sign, ignoring it.

Henri Cartier-Bresson's definition of the decisive moment as "the simultaneous recognition, in a fraction of a second, of the significance of an event as well as the precise organization of forms which gives that event its proper expression" (Friend 2004) is well understood within this context. The Israeli Arabs are mentioned in a headline above the photo which relates to the anti-Arab slogans of Liberman's party. Thus it maintains a dialogue with the range of visual images inhabiting the newspaper spread. The photo hints at two central elements in the platform of the right-wing party and of its chairman: the attitude toward Israeli Arabs—embodied in the slogan that appears on the billboard, namely, "No Citizenship without Loyalty"—and the opposition to religious coercion. Placing them in symmetrical opposition, it highlights the otherness of the Arabs, of the orthodox Jews, and of Lieberman himself as the leader of a political party with a particular political orientation in Israeli society. All these elements are present in the linguistic landscape of this random street scene, and also in the linguistic landscape consciously created by the photographer and later by the paper editors. This image demonstrates the power of the linguistic landscape to create a clash between the meaning of spaces and the meaning of words (e.g., in street signs or graffiti) which are present within their domain, causing the viewer to reevaluate them (Ben Rafael et al. 2006; Hanauer 2004; Elkins 2002; Conquergood 1997).

The Arabs, visually absent from the photograph, are nevertheless there thanks to the presence of the orthodox Jews, who are just as "other" and as "unloyal" as the Arabs, and are apparently oblivious of the campaign poster hanging above their heads. The ironic contrast in the photo stems from the familiarity of the Israeli viewer with the skepticism of the orthodox Jews as to Lieberman's ability to change their status in Israeli society. Finally, there is also a sense of self-disparagement on the part of the so-called "loyal" Israeli citizens who look at the photo and offer their own definitions of loyalty and otherness.

Israeli society employs humor to observe and condemn, to reinforce entrenched positions, and sometimes to preserve a general consensus. Usually, the humoristic aspects in Israeli press photos are not "funny" or "cute," but rather evoke a bitter smile, in the milder cases. This stems from the constant sense of temporariness, and from the intensity and effort to maintain solidarity that characterizes Israeli society. This reality affords a wide range of humorous expressions, most of them based on identifying surprising elements and presenting them in a charged context. The humor in the photos often involves self-disparagement, irony, scathing sarcasm, protest, or a sense of despair, anger, or helplessness. In the more lighthearted cases, the humor is used to celebrate the absurd, and convey a sense of surprise and self-admiration at the fact that Israel and its society still exist, despite their absurdity. The humorous photos investigate all this in the open laboratory of the street and of every-day life. Often, they aim to shake up the viewer and condemn society's indifference to the dangers inherent in the acceptance of absurd situations. In some photographs humor serves as a kind of compass, and as a call to recognize our own otherness as a preliminary condition for accepting the other (Benhabib 1992).

The two topics discussed above—the perception of humor which changes according to the channel of its expression, and the sensitivity of Israeli society to the use of humor in given situations—are both dynamic, and the changes they undergo are manifest in the complex interplay between them. The use of irony in photography, to be discussed below as a basis for a broader discussion of humor in photography, is especially relevant to the investigation of multifaceted humor as mechanism of survival in Israeli society, which is characterized by polarity and the need to tolerate extreme situations.

IRONIC EXPRESSION AND SHADES OF HUMOR IN PHOTOGRAPHY

Saying one thing while meaning the opposite, one of the basic definitions of irony, is problematic when applied to photography. As Scott (2004) puts it, how can a photograph, which is an imprint of light as it existed at a particular time and place, show one thing yet mean another? And what about the documentary aspect of photography, the "I was there" aspect, which is perceived as a defining attribute of this medium and evokes the complex dialogue it maintains with notions of memory and history

Dr. Johnson defines irony as "a mode of speech in which the meaning is contrary to the words" (Johnson 1755). This means that both the ironist (i.e., the producer of the ironic expression) and the audience must simultaneously perceive a meaning and its opposite, or else interpret some aspect of the expression as contradicting its basic meaning. This understanding is fully accomplished only when one recognizes the comic nature of the reversal, as well as the essential role of the comic factor in the full understanding of the original expression, which is being put up for criticism. Even if the act of interpretation seems almost instantaneous, it requires the addressees to carry out a gradual process: first of all to understand the face-value meaning of the (spoken or written) proposition, and then to consider the implications of the context, which yield the ironic reading. Only then can they get the "punch line" that presents the incongruity or ambiguity as amusing. This is how ironic expressions, and humorous expressions in general, can be used to imply various shades of doubt and to signal potential criticism.

Scott (2004) suggests that photographs can convey two kinds of irony: word based and situation based. She resolves the paradox of a single image simultaneously conveying two contrasting meanings by suggesting a linear interpretation model, whereby different aspects of the image are processed chronologically and analyzed with the same tools used to analyze verbal discourse or literary texts. In the second part of this chapter I will apply some of her insights in my discussion of a selection of photographs.

The investigation of ironic expression in photography as one of the uses of critical humor is also enticing in light of the epistemological parallels that exist between irony and photography. To express irony, one must point to a source, refer to it, and criticize it by reversing its meaning. The ironic expression thus compels us to tease apart and signal meanings. Consequently, it requires outlining a linear progression: a narrative that connects two conflicting aspects of form or meaning within the domain of a given photograph.

Another alluring aspect of irony is that it involves a subtle "intelligent trick" whose creation and comprehension compliments the ironist and the viewer who deciphers the message, respectively. This subtle trick allows the ironist to convey subversive messages with an innocent face, but also involves the risk of missing audiences (Muecke 1969, 1973).

Scott (2004) argues that irony is becoming more and more prevalent in art, and especially in photography, for several reasons. First, the age we are living in is concerned with shattering dominant meta-narratives and suggesting numerous alternative and subjective narratives in their stead. Simultaneously, there are many self-doubting and multi-faceted perspectives, which reflect a rejection of absolute terms, and this increases the fondness for works that have a playful aspect. This ideological climate corresponds to the epistemological state of contemporary photography, which is no longer perceived as an objective record of reality but as one possible representation of reality among many. This fundamental perception of photography affords greater scope for highlighting the fluidity of texts, each of which is a legitimate representation in itself, and stresses the equal status given to different representations of reality. It also recognizes the playful pleasure inherent in the ability to create these different worlds. In other words, the staged reality, or the world created by the artist in his studio, may be accorded the same value as a "live" press photo of a political event, for example. Even photography defined as "documentary" is no longer perceived as an objective "testimony," but only as one possible testimony. This perception affords the photographer greater creative and interpretive freedom, and widens the scope for seeing various shades of humor as a critical tool that has the gravitas of "truth" (Lucaites and Hariman 2001). The humorous moment created in the studio and the one incidentally captured by the camera may thus be perceived as equal both in their legitimacy as representations and in their complex impact potential.

Faced with a specific task, most press photographers are predisposed both internally (in terms of their professional training) and externally (by the demands of the editorial board) to look for compositions containing internal conflicts and contrasts (Nathansohn 2007; Zelizer 1995). A photo of a Jewish settler who was prosecuted for shooting a Palestinian appeared under the headline "Israeli Defense Minister Barak: 'The judicial system has been lenient with settlers who break the law'" (Figure 3.1).

The halo formed by the window behind the settler's head might have been perceived as incidental if it were not for the tight link between the headline, which calls him a "settler," and the Christian iconography that is applied here to a religious Jew. The photographer's and viewers' previous familiarity with this iconography, the left-wing political orientation of the *Ha'aretz* (in Hebrew: "The Land") daily, in which this photograph appeared, and the headline chosen by the editor create the following scheme: the surprising contrast between meanings from similar fields (religion—Judaism and Christianity—and "saintliness" as opposed to criminal charges) generates an initial sense of amusement, while the content of the headline lends condemnatory overtones to the humoristic expression. Clearly, this is an interpretation rather than a direct message of the photo, and the use of humor takes away some of its sting. Nonetheless, the systematic application of this type of humor to a certain sector will ultimately help to consolidate certain stereotypes about it.

Figure 3.1   Elections, Tess Scheflan, 2009.

32 *Ayelet Kohn*

Figure 3.2 Tel-Aviv Beach, Alex Levac, 2005.

This photograph by Alex Levac, which appeared in *Ha'aretz* and later in a book by this award-winning photographer (Levac 2008), can trigger an immediate sense of amusement, especially because we are predisposed to recognize humor in this particular photographic genre and especially in the works of this particular artist, which are familiar to Israelis from his weekly column in *Ha'aretz*.

Beyond the surprise triggered by the clash between the two couples, this photo involves a subtle expression of emotion, which replaces the banality of direct emotional stimulation or intellectual stimulation. The initial amusement is accompanied by a slight sense of unease and awareness of suppressed emotions. This understanding is afforded by the internal division in the photograph, and by the contrasts within it—between revealing and concealing, ritual and routine—which are anchored in the contrast of materials and textures (wedding clothes vs. sand), giving rise to various comments on social perceptions (nakedness, conservativeness, modesty, playfulness, and the "game of life"). A discussion of the conflicting meanings is facilitated by the natural division performed by the eye, which splits the photograph into two domains. This internal division marks the (almost imperceptible) pause performed by the viewers as they decode the image. This adds to the photograph a linear, chronological dimension, which enables the construction of a narrative with a humorous twist, a sort of "visual punch line." The understanding of this twist gives rise to further insights, also anchored in realizing the roles of the

From a Slight Smile to Scathing Sarcasm   33

humorous expression. In this case, the juxtaposing of conflicting meanings serves to index various fears and prohibitions pertaining to intimacy. The familiar frame of this easily recognizable genre allows the combination of the fear and the alleviation and defusing of this fear into one united and simultaneous "decoding-capsule."

The Role of Written Words in Creating Humorous Situations

The ironic gap, in which the readers know more than the photographed object, as a strategy of humor, determines the level of ambiguity in a given situation. This gap is created by the organization of the photograph, by the relation between the elements in the frame, and by the different orientations of the gaze. The following examples emphasize the central role of written words in inviting an ironic interpretation, or a twisted smile that is instantly aware of the criticism conveyed by the humorous expression in the visual image.

Taken in 1983 during the first Lebanon war, this is a portrait of then defense minister Ariel Sharon, by photographer Micha Kirshner. It was published in Kirshner's book *The Israelis* (1997), along with a text by journalist Ron Meiberg describing the circumstances of its making:

> "Nobody move," Kirshner orders Sharon—although, according to Sharon's detractors, following orders is not one of his more prominent personality traits. "I still don't have a photograph." The brutality of

Figure 3.3   Ariel Sharon, Micha Kirshner, 1983.

34 *Ayelet Kohn*

> his tone is enough to glue [Sharon's advisor] Uri Dan to the wall. "Get up, please," Kirshner commands Sharon, "and stand against the wall. Yes, with your back to this wall, and don't move. Don't look at me and don't smile." [Journalist Ehud] Yeari and I watch with amazement as Kirshner positions Sharon next to a large frame on the wall, which contains the utopian [Biblical] verse "They will beat their swords into plowshares and their spears into pruning hooks", a gift to Sharon from [his wife] Lily.
>
> Sharon stands next to the verse, as commanded by Kirshner, whose Nikon [camera begins to] shoot bursts—at least two rolls of 35 mm film. Uri Dan stands by, shell shocked. (Kirshner 1997:58–61).

The staging of this scene intensifies the conflict between the Biblical verse (Isaiah 2:4), highly familiar to Israelis from various national ceremonies, and the figure of Ariel Sharon. The effect is heightened by the use of two elements: the decisive moment genre (the text juxtaposed with the long-nosed shadow which evokes a demonic caricature) and the technological means, namely, the distorting lens. The sense of irony stems from Sharon's obliviousness to the effect created by the whole, especially because the text—a gift from his wife—was important to him, which is why he asked to be photographed against it. As in the other examples, here too the ironic message clearly depends on the context, on the photographer's stance, and on the viewers' ability to identify all its components and allusions (see also Mitchell 1994).

The double clash of meanings can manifest in either content or form, and we take an interest in photos that involve multiple points of contrast, and in the ways in which humor utilizes these contrasts to illuminate existential questions. The situation under discussion raises questions regarding the nature of the clash and its effect. The differential knowledge allows the photographer and the journalists to express a political position, and the technical means create a caricature-like effect. Like the anti-Semitic caricature, this image is not intended to evoke a smile but to express criticism by making Sharon seem ludicrous. Sharon's obliviousness allows the photographer to cast doubt upon the motivations of the then-defense minister. He is not just making fun of Sharon, but presenting him as an actual threat.

Shades of Humor Depending on Situation

Edgar and Sedgwick (2007) note that, in addition to creating a "community of comprehenders," irony also allows the speaker to remain distant from what he or she is discussing and to focus on the shaping of the outward shell, namely, the humorous frame of expression. The previous example involved humor in a photo containing written text. The next one involves humor based on comprehension of the situation and the different degrees of involvement of the photographers who identify the ironical situation and relate to it.

The same is true for a photo by Pavel Wolberg, portraying a couple during the Tel-Aviv Gay Parade. The man is wearing a tutu skirt and carries a gun over his shoulder. The woman is dressed as a dancer. Here too the joke is twofold, because the sense of irony is triggered not just by the alienness of the costume and the contrast between the masculine body and the feminine tutu skirt, but mainly by our familiarity with Israeli reality, where a civilian carrying a gun is not a rare sight. On second thought, the strange element in the photograph is not the skirt on the man's muscular body, but the gun slung across his back, which, on the one hand, makes him seem even more ludicrous, but at the same time also directs a critical remark at what is perceived in Israel as commonplace.

Adopting Palmer's four-stage model for understanding the relationship between the components of a humorous expression, we can propose the following analysis, which lays out the process of response to the photograph, from the initial sense of amusement to the ultimate comprehension of the critical message. The initial sense of surprise is triggered by the image of a man in a skirt, a spectacle not frequently encountered on an Israeli street. However, a logical explanation is immediately provided by the context: a parade where people wear costumes (the Pride Parade, or perhaps a Purim carnival). This comprehension makes the abnormal seem normal. The proposition "a man in a skirt is different from a real man" thus becomes "a man in a costume is a common sight in a particular context." Another proposition accepted in Israeli society is "men carrying weapons are a common spectacle." A uniformed soldier carrying a gun, or even an armed civilian, may be a surprising sight to a tourist, but Israelis would barely notice it. According to Palmer's model, it is the combination "a man in a tutu skirt carrying a gun" (slung rakishly across his back) that leads the decoder to a new comprehension based on the one preceding it. Once the anomalous phenomenon (man in a tutu skirt) is explained and comprehended, our sense of doubt is directed toward the proposition that is apparently accepted and unsurprising, namely, armed civilians are a common sight.

This photo thus utilizes a situation immediately identified as comical to make a social comment. This comment arises out of the linear progression from surprise, to discontinuity, inconsistency and a feeling of unease. In this sense the photo can be seen as transgressive (Nathansohn 2007; Foucault 1977).

The Gaze as a Means of Marking the Implausible

A second element which, I suggest, serves to highlight implausible situations, and to illuminate the absurdity created by conflicting readings of a given photo, is the aspect of the gaze, whose key role must be analyzed and understood. The gaze singles out incongruous aspects of a given situation and puts them up for critical examination. The absurdity of the situation as a whole, or one of the figures in it, can be manifest in either content or form. Shooting angles and distorting lenses, for example, can emphasize flaws or conflicts in a photograph, thereby illuminating some reality as problematic. The gaze thus serves to signal incongruity, thereby taking humor to the caustic end of its spectrum, where it transforms into scathing sarcasm. This particular shade of bitter humor is immediately understood to be far removed from lighthearted comedy. It is not a stimulus that triggers a reflexive and limited response of amusement, but is closer to harsh satire and self-criticism that is aimed at exposing injustice.

By examining the relations between the gaze of the photographer, the gaze of the viewer, and the presence of the gaze or gazes within the photograph, we can show that this element plays a key role in constructing an image where the surprising contrast between components presents the situation as a whole, or one of the figures within it, as absurd, with the aim of conveying political criticism. I will suggest further that the gaze can resolve the dilemma posed by the tension that exists in any photo between its synchronic and diachronic interpretations. A contemptuous gaze can sharpen the critical attitude toward elements by reversing them, and can even cast doubt on the "truth" represented by each visual image. The gaze functions as an element that introduces a chronological ordering into the synchronic text of the photo, where the comprehension of the incongruous and exaggerated situation leads to critical self-examination.

Paraphrasing Mulvey's (1975, 1989) concept of the "penetrating gaze," and associating it with the notion of the search for narrative triggered by a figure moving through space, I suggest to call the gaze in question a "cross-eyed gaze." It is a direct look which achieves its goal by focusing on a message that exists on the periphery of the photo, or by directing the viewer's attention out of the photo itself and into the photo's blind area, which is ostensibly unfocused (Barthes 1981). This gaze conveys criticism due to the deliberate reversal inherent in it. The viewer who looks, peeps, gazes, or stares at the scene notices "extra meanings" that "leak" out of the deliberately constructed framework. This leads us to question how the photographer sets up the situation so that the periphery takes precedence to the center, and so that the gaze, as a marker of incongruity, highlights the internal clash and directs the viewers toward an interpretation of bitter contempt.

In Israeli reality, extreme situations are perceived and experienced as an existential necessity and an integral part of the country's essence. Like Israeli discourse, this reality is brutal and aggressive (Katriel 1986, 2004), and humor—especially self-directed and mostly critical humor—serves as a means of tolerating and surviving it. Living in a state of ongoing conflict means living with a paradox: normality is perceived as a state of equilibrium within a complex system of conflicting factors, and the daily negotiation over ways of coexistence is a major preoccupation in all arenas of life, ideological and practical, individual and collective.

This understanding is natural to any photographer who wishes to "tell a story." It is certainly natural to a photographer who works in Israel and

From a Slight Smile to Scathing Sarcasm 37

is well aware of this existential paradox. Therefore, many photos taken in Israel involve two levels of contrast. First, contrast that presents "conflict as the norm." These photographs are taken in environments offering a photogenic and easily identifiable conflict, for example, Jewish settlements in the occupied territories, the Knesset (Israeli parliament), Arab villages on Israeli memorial days, and so on. Experience has taught the photographers that these photographs sell well (Zandberg and Neiger 2005; Nathansohn 2007; Ilan 2008; Ewha 2003). The second level is the construction of conflict within the image by contrast, lighting, a unique feature and design of the composition within the frame.

In discussing the gaze, we ask ourselves who is peeking at whom at any given moment, and wonder about relations of ambush, surveillance, and a gaze that reveals and conceals, exposes, and mirrors other gazes in an actual "decisive moment." In Figure 3.4, by Pavel Wolberg (Tufah village), Paul Frosh's (2001) notion of "scopic regime" takes on a political meaning anchored in comprehending the absurdity of the situation. This breakfast scene was photographed in the home of a Palestinian family in Tufah, a village near Hebron. We can read it as a condemnation of the "normalcy" of depicted routine, and as criticism of its absurdity, but also as a comment on the humanity of both sides.

The orientation of the gaze—both the placing of the camera and the location of those who look at it or ignore it—redefine the power balance within the photograph (Jay 2002). The eye of the photographer, which subordinates his hand, along with the hand of the soldier which holds the gun,

Figure 3.4 Tufah village, Pavel Wolberg (printed in Wolberg, 2006).

supervise and control the scene, and may also contaminate it and even kill. Thus, the gaze that controls the visible body itself becomes an object of ideological commentary (Synnott 1992).

Here, the photographer's gaze is not a simple gaze that only looks but does not appropriate what it sees (Merleau-Ponty 1974). In a photograph like this, the camera is an apparatus that models the thinking process and serves an epistemic function (of changing the significance of the world). Should the incongruous balance of the photograph be upset, the camera will witness the moment at which the weapon becomes a tool that also serves a transformative function (of changing the world itself) (Douglas 1975).

This frozen moment evokes the sense of violence that is inherent in any act of photography and also in the act of invading someone's home. The eyes of the soldier, of the family, of the photographer, and of the viewer all meet at a point that marks the situation as alien, a place where gazes refrain from meeting. The decision to use a wide angle while focusing on the soldier, which is familiar to the viewers as a technique of ridiculing, directs the surprise toward a sense of criticism. The distorting lens, the soldier's expression, and the presence of the camera emphasize the absurdity of the situation and the varied levels of recognition and avoidance it raises.

CONCLUSION

This chapter examined a few methods through which political and social photography in Israel utilizes shades of humor and as a means of criticizing and challenge dominant and iconic representations such as national emblems and occupied territories. Like satire and parody, humorous photography can present the reality it depicts as authentic and at the same time doubt or condemn it.

Theoreticians in the field of photography speak of ambiguous irony. The discussion of the photographs in this paper touched on the key role of the discrepancy between the knowledge of the person in the photo (or the system she or he represents), who is the butt of the joke (benign or contemptuous), and the knowledge available to the viewers. This discrepancy (like the ironic gap in tragedy) focuses the viewers' attention on faults and injustices, which the media often presents in a vague or banal manner that numbs sensibilities. The ways in which this gap is exhibited and displayed are determined by the organization of the photo, the construction of the conflict, the presentation of the relationship between the elements within the frame, and the orientation of the gaze. The discussion of the photo's internal composition, which allows a gradual communication and decoding of the humorous message, illuminated the role of the gaze—both the photographer's gaze and other gazes within the photograph—in criticizing what is usually perceived as a mundane and even normal situation.

The absence of the gaze—manifest in choosing to show figures from the back, or to depict figures who are unaware of something occurring behind them—shed light on the power of the gaze that is present in the photograph, on the one hand, and on the criticism inherent in presenting the object as vulnerable and as the focus of the ironic gap, on the other. In these cases, attention is focused on an event of which the photograph's protagonist is oblivious. The viewers are invited to notice this vulnerability, the ironic gap, and the power of the gaze that juxtaposes the conflicting elements in the photograph. In the next stage of the decoding process, they are invited to understand the gaze that is directed upon them and requires them to perform an act of self-examination.

These political photos which appear in the Israeli press, either by themselves or accompanied by text, utilize shades of humor to challenge norms and express criticism. They are related to genres that make condemnatory use of radical and humorous representations (from caricature and grotesque to satire and parody), whose social, ideological and political purpose is to connect amusement to surprise, and surprise to shame and self-criticism. The "testimony" factor, so central to photography, is also present as one deciphers the humorous message: the viewers are invited to serve as embarrassed witnesses who are themselves subject to the critical "cross-eyed penetrating gaze" of the press photographer.

NOTES

1. Cartier-Bresson's notion of capturing a precise moment which brings together different features of the composition to create a surprising, fresh conception of the depicted situation.
2. Alex Levac is a winner of the Israel Prize (2005), and Micha Kirshner is head of the photography department at WIZO College in Haifa. Some of their works, as well as those of Pavel Wolberg and Micha Kirshner, were exhibited in art galleries and museums.
3. This slogan reflects Lieberman's suggestion to deny citizenship to Arabs unless they pledge loyalty to the State of Israel.

REFERENCES

Barthes, R. 1981. *Camera lucida: reflections on photography*. Translated by Richard Howard. New York: Hill & Wang.
Ben Rafael, Eliezer, Elana Shohamy, Mohamad Amara, and Nira Trumper-Hecht. 2006. "Linguistic landscape as a symbolic construction of the public space: the case of Israel." *International Journal of Multilingualism* 3(1):7–31.
Berger, John. 2003. "Photographs of agony." P. 289 in *The photography reader*, edited by L. Wells. London: Routledge.
Conquergood, Dweight. 1997. "Street literacy." Pp. 354–375 in *Research on teaching literacy through the communicative and visual arts*, edited by J. Flood, B. H. Heath, and D. Lapp. New York: Macmillan.
Douglas, Mary. 1975. Implicit meanings: *essays in anthropology*. London: Routledge & Kegan Paul.

Edgar, Andrew and Peter R. Sedgwick, eds. 1999. "Irony." In *Key concepts in cultural theory*. London: Routledge, pp. 132–133
Elkins, Jay. 2002. "Preface to the book *A skeptical introduction to visual culture*." *Journal of Visual Culture* 1:93–99.
Friend, David. 2004. "Cartier-Bresson's decisive moment." *The Digital Journalist*, December.
Frosh, Paul. 2001. "The public eye and the citizen voyeur: photography as performance of power." *Social Semiotics* 11(1):43–59.
Hanauer, David I. 2004. "Silence, voice and erasure: psychological embodiment in graffiti at the site of Prime Minister Rabin's assassination." *The Arts in Psychotherapy* 31:29–35.
Ilan, Yonatan. 2008. "Shooting news photos—showing the world: processes of producing news photos in international news agencies." MA thesis supervised by Paul Frosh, Department of Communications, Hebrew University of Jerusalem (Hebrew).
Jay, Martin. 2002. "That visual turn." *Journal of Visual Culture* 1(1):87–92
Jewitt, Carey and Rumiko Oyama. 2002. "Visual meaning: a social semiotic approach." Pp. 134–156 in *Handbook of visual analysis*, edited by T. van Leeuwen and C. Jewitt. London: Sage.
Johnson, Samuel. 1755. "Irony." *A dictionary of the English language.* http://archive.org/stream/dictionaryofengl01johnuoft#page/n5/mode/2up
Katriel, Tamar. 1986. *Talking straight: "dugri" speech in Israeli Sabra culture.* Cambridge: Cambridge University Press.
Katriel, Tamar. 2004. "Talking straight: the rise and fall of dugri speech." In *Dialogic moments: From Soul Talks to Talk Radio in Israeli Culture. .* Detroit, MI: Wayne State University Press.
Kirshenblatt-Gimblett, Barbara. "Traditional Storytelling in the Toronto Jewish Community: A Study in Performance and Creativity in an Immigrant Culture." Dissertation. Indiana University, 1972.
Kirshner, Micha. 1997. *The Israelis*. Tel-Aviv: Hed Artzi.
Levac., Alex. 2008. *Israel: the twenty-first century.* Jerusalem: Carmel and Libros.
Lister, Martin and Liz Wells. 2002. "Seeing beyond belief: cultural studies as an approach to analyzing the visual." Pp.61–91 in *Handbook of visual analysis*, edited by T. van Leeuwen and C. Jewitt. London: Sage:
Lucaites John Louis and Robert Hariman. 2001. "Visual rhetoric, photojournalism, and democratic public culture." *Rhetoric Review* 20(1–2): 37–42.
Mitchell, W. J. T. 1994. *Photograph theory: essays on verbal and visual representation*. Chicago: University of Chicago Press
Muecke, Douglas Colin. 1969. *The compass of irony*. London: Methuen.
Muecke, Douglas Colin. 1973. *Irony*. London: Methuen.
Mulvey, Laura. 1975. "Visual pleasure and narrative cinema." *Screen* 16(3):6–18. Online version.
Mulvey, Laura. 1989. *Visual and other pleasures*. Bloomington: Indiana University Press.
Nathansohn, Regev. 2007. "Shooting occupation: the sociology of visual representation." *Theory and Criticism* 31:127–154 (in Hebrew).
Nielsen, Alleen Pace. 1986. "We should laugh so long?" *School Library Journal* 33:30–34.
Palmer, Jerry. 1987. *The logic of the absurd*. London: BFI Publishing.
Scott, Biljana. 2004. "Picturing irony: The subversive power of photography." *Visual Communication* 3(1):31–59.
Shifman, Limor. 2007. "Humor in the age of digital reproduction: continuity and change in Internet-based comic texts." *International Journal of Communication* 1:187–209.

Sperber, Dan and Deirdre Wilson. 1990. "Rhetoric and relevance." Pp. 140–156 in *The ends of rhetoric: history, theory, practice*, edited by J. Bender and D. Welbery. Stanford, CA: Stanford University Press.

Sperber, Dan and Deirdre Wilson. 1991 [1981]. Irony and the use–mention distinctions. Pp. 550–563 in *Pragmatics: a reader*, edited by S. Davis. Oxford: Oxford University Press.

Synnott, Anthony. 1992. "Tomb, temple, machine and self: the social construction of the body." *The British Journal of Sociology* 43(1):79–100.

Wolberg, Pavel. 2006. Tel-Aviv: Dvir Gallery.

Zandberg, Eyal and Motti Neiger. 2005. "Between the nation and the profession: journalists as members of contradicting communities." *Media, Culture and Society* 27(1):131–141.

Zelizer, Barbie. 1995. "Journalist's 'last' stand: wirephoto and the discourse of resistance." *Journal of Communication* 45(2):78–92.

Zuromskis, Catherine. 2008. "Outside art: exhibiting snapshot photography." *American Quarterly* 60(2):425–441

4 Sociology of Iconoclasm
Distrust of Visuality in the Digital Age

Łukasz Rogowski

Visual sphere as an object of sociological study may be understood in several ways. The first one defines pictures as messages and thus interprets them as an element of communication processes. This procedure—referring for example to the semiological or content analysis (Rose 2001:54–99)—is based on the assumption that pictures are created in social contexts and that their meanings are "mirrors of the society." It is always the participants of social life who interpret a picture, and their interpretation is based on social norms, values, or beliefs; that is why it may be the subject of sociological research. Second, the broader context of the creation, production, and usage of pictures may be taken into account. A sociologist may research such topics as the materiality of pictures (Edwards and Hart 2004), the relationship between pictures and memory or identity (Burgin 1996; Hirsch 1997), and the role of pictures in (re)producing social inequalities and social ties (Bourdieu 1984[1979], 1990[1965]). The third way of understanding the visual sphere goes beyond the pictures. In this context, sociology may be interested in, for example, visual aspects of social interactions (Goffman 1963, 1971), social rules of usage of the sense of sight (Garland-Thomson 2009), mental images (Mitchell 1984; Forrester 2000), or axionormative systems concerning the role of visuality in social life and its historical variation (Lalvani 1996; Crary 1999). Visual sociology refers also to the methodological aspect of social research, but this will not be discussed in details in this chapter.

In this chapter[1] I would like to refer to the third way of understanding the visual sphere: as a pretext for researching other aspects of social life and answering the most important questions of sociology, such as social ties, social inequalities, or social change. The purpose of the paper is to describe selected sociological aspects of iconoclasm. As this phenomenon is usually associated with religion, I will start from describing the most important religious regulations of the visual sphere. However, in further parts of the chapter I will extend the range of my interests and refer not only to the religious aspects of iconoclasm, but also its political or technological aspects. I concentrate especially on contemporary varieties of iconoclasm, trying to answer the question how the transformation of modern societies and technologies affect the social usages of visuality.

Iconoclasm is usually defined as a physical destruction of material pictures[2] (van Asselt et al. 2007:1–29; Bryer and Herrin 1977). However, as sociologists should research not only actions, but also the attitudes and opinions, I use a wider definition of iconoclasm, which refers to any distrust of visuality. Such an approach follows, for example, Michel Maffesoli (1996[1988]:137) or David Freedberg (1989), and allows to describe not only the destruction of pictures, but also its socio-cultural causes and effects. It is helpful also in the studies of digital pictures and the sense of sight.

ICONOCLASM AS A CONCEPT AND A SOCIO-HISTORICAL PHENOMENON

Distrust of visuality is not a homogeneous phenomenon. First, it may result from different causes. It may be based on individual beliefs, previous experiences, or even physiological or cognitive features (e.g., blind people's attitudes to visuality). But on the other hand, it may be characteristic not only of individuals, but also of whole communities, societies, or cultures. In this situation, an attitude toward visuality is a result of group's historical, geopolitical, or economic situation. Sometimes it also refers to legal regulations. For example, the Old Testamentary regulations regarding visuality, which I describe later in this paper, were the result of the geopolitical situation at the time and of the desire to distinguish the people of Israel from other tribes which produced idols.

Second, the consequences of distrust may also be very diverse. The most obvious is the destruction of material pictures (but also paintings, sculptures, monuments, etc.). This results from a negative attitude to visuality and includes not only cognitive but also emotional response, as well as behavioral intention. But defining visuality as a "worse form of communication," which I discuss later, following Bruno Latour (2002), may take effect in less radical consequences. The examples could be visual or media literacy (Messaris 1994; Tyner 1998), which propose educational programs teaching responsible usage of visuality and immunity from visual persuasion or propaganda. Sometimes, the reception of visuality may be regarded as requiring certain prerequisites: knowledge, skills, competences, but also an appropriate age or gender. It is considered necessary to impose restrictions on the availability of pictures (the examples may be age limits for movies or websites).

As Freedberg (1989:378–428) claims, iconoclasm can be explained by referring either to the ontological status of pictures or to the social functions of iconoclastic practices. Whereas the ontological status of pictures explains the causes of iconoclasm, the social functions of iconoclasm refer to its consequences. The former focuses on the process of representation (Hall 2000), which assumes some kind of correspondence between the signifier (in this case: the picture) and the signified (what is in the picture). For example, a visual representation of a "cat" (an animal) occurs when a picture of a cat (signifier) contains all the most important features of the idea of a cat (signified). Iconoclastic practices occur when someone notices a discrepancy between pictures and their mental or ideal prototype—a picture may then be considered "not real," "false," or "far from its prototype." This may lead to the viewer's desire to destroy the picture, which may be based on different motivations. This kind of iconoclasm was described by Bruno Latour (2002), who distinguishes between five possible types of actions aimed at visuality. He uses consecutive letters of the alphabet to name various types of people who participate in these actions. A-People are those who are opposed to all pictures, without exception, as they perceive them as obstacles in the pursuit of the truth. B-People do not object to the pictures themselves, but rather to their imperfection. Latour considers himself a member of this group. Pictures are important for him only if they help to seek the truth. That is why we should continuously try to produce better pictures. For C-People, pictures are important only if they are symbols of an opposing group. A flag, a banner, or an emblem is not valuable as such, but its seizure and/or destruction is a symbolic demonstration of domination or superiority. D-People are those who unconsciously destroy pictures. Finally, E-People are "simply the people" whose activities toward pictures are only a form of temporary entertainment.

Focusing on the social functions of iconoclastic practices, we may ask not only why people, groups, or institutions distrust or destroy pictures, but also what for. As for the latter question, the most general answer may be that this is an attempt to maintain control over different aspects of individual and social life, in the fear of losing it. First, the iconoclasm attempts to standardize the social life and its manifestations (Turner and Turner 1978:140–143). For example, when invaders conquer new territories and incorporate them into their country, they also try to eliminate local pictures referring to the history of the territory or the nation. Their purpose is to destroy social memory and to clarify the new structure of the state (Gamboni 1997:29). This case of iconoclasm is the reaction of a government and/or the dominant class to some kind of behavior that may threaten their reign and/or social cohesion. They try to standardize imagery—and thus destroy its local varieties—in order to legitimize their dominance. Second, iconoclasm creates possible forms of manifesting the fear of picture itself and of the emancipation of visuality from its creators. As William J. T. Mitchell (2005) states, pictures are the rightful actors of social life. They are alive because they have desires, they love or hate, and thus they may influence people's activities. Iconoclastic practices are aimed to ensure that the picture will not begin to live its own life and will not break the predictable rules of social life. For example, Moses destroyed the Golden Calf, created by the people of Israel, because he regarded it as a visual statue inducing to violate the God's law. Third, iconoclasm allows one to control "otherness." Destroying pictures of the other (e.g., tearing pictures) induces a magical belief that the destruction also influences a person or a group of people represented in the picture. But the same damage can be as well a display of different opinions and an attempt to manifest one's own values. To sum up, iconoclasm can refer both to macro- and microstructural aspects of social life. On one hand, it is a way of confirming and legitimizing one's own power; on the other, it allows to control your own activities and sensuality.

In the historical context, the concept of iconoclasm is most frequently described in relation to religion, which regulated the basic principles of social life in traditional, pre-modern societies. It created legal and customary rules, which determined socio-cultural practices, including those related to visuality. The best examples are biblical principles. One of the Ten Commandments, given by the Lord to Moses on Mount Sinai, is, "Thou shalt have no other gods before me. Thou shalt not make unto thee any graven image, or any likeness of anything that is in heaven above, or that is in the earth beneath, or that is in the water under the earth. Thou shalt not bow down thyself to them, nor serve them" (Exodus 20, 3–5[3]). Visuality is regarded as a potential opponent of "the true worship," whereas iconoclasm does not refer in this case only to the destruction of pictures, but more broadly to the prohibition of their creation. It also applies to the rules of sight:

> And he said, Thou canst not see my face: for there shall no man see me, and live. And the Lord said, Behold, there is a place by me, and thou shalt stand upon a rock: And it shall come to pass, while my glory passeth by, that I will put thee in a clift of the rock, and will cover thee with my hand while I pass by: And I will take away mine hand, and thou shalt see my back parts: but my face shall not be seen. (Exodus 33, 20–23)

Defining visuality as a worse medium of communication and worship partly changed with the rise of Christianity. The Old Testamentary rules, which were still obligatory for first Christians, had to be compared with the doctrine of the Incarnation, which defines Jesus Christ as the visible God and man. So, if it was possible to see Him, it also should be possible to produce pictures of Him. In AD 313 the Emperor Constantine the Great issued the Edict of Milan, which introduced religious tolerance in the Roman Empire and thus included Christian worship in the public sphere. Since then the Christian art—including both western and eastern traditions—has become one of the basic elements of visual arts. Before the Edict was issued, the worship and therefore its visuality had been forced to be in hiding.

But from the very beginning of Christianity there were opinions depreciating the value of visuality in religious life. Pope Gregory the Great (c. 540–610) described pictures as "the Bible for those who cannot read." Hereby he sanctioned the presence of visuality in Christianity, but also pointed to subordination of picture to word and language. This has led to the coining of the Latin term *biblia pauperum* ("the bible of the poor") to depict visual forms of Christian preaching such as illustrated books, paintings or stained-glass windows, used by illiterate people who could not read the Bible[4] (Carpo 2001[1998]:81–88).

Controversies over visuality resulted in two iconoclastic revolutions in Christianity: the eastern (orthodox) in the eighth century and the western (protestant) in the sixteenth century. The former referred to the theological and philosophical questions about the possibility of visual representation of God. The latter was a part of the Reformation, a religious (but also political) movement challenging the authority of the Pope and the Catholic tradition. In this case, removal or destruction of pictures was the most apparent manifestation of the Reformation in everyday life. These two revolutions resulted from different assumptions and different aspects of iconoclasm mentioned by Freedberg (1989:378–428). The eastern iconoclastic revolution was based on the question about ontological status of the picture, whereas the western referred to the social functions of iconoclasm.

CONTEMPORARY ICONOCLASM

Having presented the historical and theoretical aspects of iconoclasm, now I would like to describe its contemporary varieties at the beginning of the twenty-first century. Important shifts in contemporary social life affect also iconoclasm. There were three phenomena which resulted in the creation of new forms of iconoclasm. The first is the rapid socio-political change, occurring mostly in Central and Eastern Europe and the Middle East. Second, it is the individualization of social life, a characteristic of postmodernism, which contributes to the increased importance of everyday life in sociological analysis (Giddens 1991; Jacobsen 2009). Third, it is the development of digital media which has resulted in the facilitation of amateur production and dissemination of different messages on the Internet, including digital pictures (Jenkins 2006; Manovich 2001). The counterparts of these changes are three contemporary types of iconoclasm presented in this paper: transformation iconoclasm, everyday iconoclasm, and digital iconoclasm. Obviously, they do not constitute a full typology of contemporary iconoclasm. Rather, they illustrate the evolution of iconoclasm and give an insight into the constantly changing socio-cultural status of visuality.

The first type of contemporary iconoclasm is transformation iconoclasm. Destruction of pictures during rapid socio-political changes and/ or opposition to them has already happened in the past. An example can be the Protestant Reformation (mentioned above) or the Storming of the Bastille during the French Revolution in 1789. However, the system transformation in Central and Eastern Europe induced by the collapse of the communist system and the Soviet bloc in the late 1980s, emphasized the importance of transformation iconoclasm and carried it to a new level, previously unknown on such a mass scale. What's more, the development of mass media enabled to inform multiple recipients all over the world about these events.

These phenomena are usually described in the political and economical contexts, but their visual aspects are also worthy of our attention. Socio-political changes involve destruction of the symbols of the bygone era. One of the key symbols of Europe divided into two hostile blocs was undoubtedly the Berlin Wall. After the Second World War, German territory, including Berlin, was divided into four occupation zones: American, British, French, and Soviet. The Berlin Wall was built on the night between August 12 and 13, 1961, as a result of the increasing conflict between the former Allies. The Wall as a system of fortifications, separating Eastern (German Democratic Republic) and Western (Federal Republic of Germany) Berlin, was an over 140-kilometer-long boundary preventing movement between the sectors of the city. It also formed a mental barrier that separated the nation and the continent. The physical division of the city led to the separation of families and friends. Illegal attempts to cross the wall resulted in numerous arrests and even deaths of the refugees. It is estimated that between 1961 and 1989 thousands of people tried to escape from the Eastern Berlin. Between 100 and 200 of them were killed (Major 2010:147).

Therefore, it is not surprising that the destruction of the Wall on the night of the November 9, 1989, is still regarded as the symbolic moment of the collapse of communism. The enthusiasm of the participants of this event stemmed not only from the possibility of crossing the borders separating the city. It can also be interpreted as a form of aggression against the hated regime. The destruction of the material (thus, also the visual) symbol of this system was a form of redress of grievances. Its destruction was also a form of homage to the victims.

Similar events, albeit of a smaller scale, took place in almost every country of the former Soviet bloc. These events involved mainly destroying statues and monuments of the former leaders and prominent members of local communist parties. For example, the statue of Felix Dzerzhinsky[5] was demolished in Warsaw; in Bucharest it was Lenin's monument which was demolished; and in Tirana, Enver Hoxha's[6] monument. Earlier, during the uprising in Hungary in October 1956, the statue of Stalin was toppled. All these events were accompanied by enthusiastic crowds who regarded the destruction of the statue of the hated leader as a form of liberation from oppressive regimes. There are also other phenomena related to socio-political transformations, which may be defined as forms of transformation iconoclasm, such as social campaigns for renaming streets or changing school patrons. The goal of such campaigns as expressed in the former Soviet bloc was to eliminate the names which alluded to communist past. For example, in Warsaw, Karol Świerczewski Street was renamed Solidarity Street, Marceli Nowotko Street was renamed General Władysław Anders

Street, and Felix Dzerzhinsky Square (where the statue of Dzerzhinsky was placed) was renamed Bank Square.[7]

Therefore, transformation iconoclasm can be defined as a process of destruction of pictures which becomes a symbolic turning point in history and designates a transition between different epochs. In this case, it is always a sign or indicator of macrostructural changes and is directly related to them. That is why transformation iconoclasm is almost always an effect of events that have occurred earlier.

Another type of contemporary iconoclasm, everyday iconoclasm, is based on different assumptions. In this case, contrary to transformation iconoclasm, the actions taken at the microstructural level are not directly connected with macrostructural determinants. They may be the result of either emotions or personal experiences associated with the picture.

Destroying private or family photographs is the first example of the everyday iconoclasm. As it is often related to the end of a relationship (e.g., when a divorce or death of a spouse results in tearing the wedding photos) or to the desire to underscore the decision of breaking social ties, its functions are paradoxically similar to those of transformation iconoclasm. It also sets the cut-off point from the past and starts a "new life" under the new rules. However, there are at least three differences from transformation iconoclasm. First, this action is made in sadness rather than euphoria characteristic of transformation iconoclasm. Second, it is not a mass, collective experience, but rather an individual one. Third, as it is usually based on the desire to erase memories, it is not visually recorded on other pictures or videos.

Other examples of everyday iconoclasm can be found in pictures uploaded to the database of *Niewidzialne Miasto* ("The Invisible City")[8]. The Invisible City is a Polish-wide, sociological, and photographical project. It aims to study—through photographs—the social activities in the urban space that can be defined as "everyday urbanism" (Chase, Kaliski, and Crawford 1999) and which show the other, often unnoticed side of urban life. Some of the photos are the examples of individual opposition to certain politicians and their decisions. First, they depict various types of inscriptions on the walls, which are intended to communicate a message (often an anarchist message). Simultaneously, they "disturb" the architecture of the city, as well as its policy and planning. In this case, the destruction is not aimed at the picture itself, but rather it treats it as a platform or medium whereby a resistance to political system can be expressed. Second, they illustrate creative modifications of election posters by using their content and/or adding another message to it. As a result, the message which was intended by the poster's authors is alternated. For example, three letters were added to the word "PiS"—an abbreviated name of one of the biggest Polish political parties, *Prawo i Sprawiedliwość* ("Law and Justice"). As a result, a vulgar word *pisior* ("dick" in English) was formed. Someone else wrote a poem in the poster of Donald Tusk, the leader of the main opposition party at the time: *Kto głosuje na Tuska, temu w gardle stanie kluska* ("For Tusk you vote, on noodle you choke"). These are only a few examples from Poland, but I am sure that the readers may find similar examples in their own countries.

It is important to distinguish everyday iconoclasm from vandalism. The destruction of pictures or changing their content and meaning do not stem solely from a desire to destroy. Rather, they involve further reflection and/or wish to communicate one's own message. That is why everyday iconoclasm can be analyzed in terms of the concepts proposed by Michel de Certeau (1984[1980]). The author divides urban activities into two types: strategies and tactics. Strategies are "official actions" taken by individuals and/or institutions with power. Creating and displaying an election poster is therefore an example of strategy. However, tactics are "actions of the weak"—those who do not have power. They have no opportunity to take actions with their own rules (e.g., they have no money for their own poster campaign). Thus, they have to use the effects of the strategies already in use. Everyday iconoclasm is an example of tactics.

Everyday iconoclasm, therefore, is similar to subvertising. The latter consists of adding new elements to visual messages, mostly to commercial or political advertisements, which give them new meanings. Subvertisers' activities may be interpreted as a part of the alter-globalization movement. Sometimes, they follow the ideas of the Situationist International—a social and artistic movement aimed at the liberation of modern societies from commodity fetishism (Debord 1994[1967]). However, there are at least two important differences between the everyday iconoclasm and subvertising. First, this type of iconoclasm is rarely organized. It relates rather to the actions of protesting individuals, whereas subversive transformations of visual messages are often reproduced in various media and in different parts of the world. There are also organizations that associate subvertisers (such as Adbusters) and media networks, which often promote the results of subvertising activities (like Indymedia—Independent Media Center[9]). Second, everyday iconoclasm is not supported by any justifying common ideology. Therefore, it is not characteristic only of left-wing movements, as is the case of subvertising.

The third type of contemporary iconoclasm is the digital iconoclasm. It is based on the digital nature of new media, which opens up new opportunities to modify a picture. What may serve as examples of digital iconoclasm are pictures showing destruction. Digital technology facilitates the visual creation of events that have not taken place in the real world. It is particularly used in disaster movies. The best example is Roland Emmerich's movie "2012," the popularity of which was largely due to the advanced special effects showing the destruction caused by the end of the world. Digital creation of damage is also popular within non-institutionalized activities, for instance, when Internet users employ the features of digital technology to create, for example, videos showing the destruction of the Statue of Liberty in New York. Similarly, in 2007, a fake atomic mushroom cloud appeared on Czech TV during live coverage weather forecast.[10] Later it turned out to be the work of a guerilla artistic action.

Another example of digital iconoclasm is so-called *spoof sites*—websites that are visually modeled after the official websites of people, institutions, or companies but whose visual content is inverted. There are various websites or e-mails which pretend to be, for example, bank sites (or messages from a bank) trying to extort access codes from Internet users. There are also parodic websites, ridiculing or embarrassing their prototype. For example, whitehouse.org[11] is a parody of the official White House website,[12] whereas fuckhotmail.com[13] was a parody of the official Hotmail website.[14]

The examples of movies and spoof sites show the complexity and ambiguity of digital pictures and result in the need to modify the previously proposed definition of iconoclasm. Paradoxically, digital iconoclasm may result in striving to improve the quality of a picture. Obviously, disaster and science fiction movies were produced also in the past. However, the development of digital technology allows more detailed interference in visual communication. But this may result in undermining the faith in pictures' reality and authenticity. Thus, digital iconoclasm shows that contemporary (dis)trust of visuality is based not on the quality of pictures, but rather on their social usages and functions. Because a digital picture can be saved in many copies and recreated, this type of iconoclasm is seen as reversible and hence less destructive.

SENSUAL ICONOCLASM

So far I have described the examples of iconoclasm which concern both material and digital pictures. However, it is worth referring to another manifestation of iconoclasm, which involves not only pictures but also the practices of looking. The dominance of vision is a modern phenomenon (Jay 1998). While ocularcentrism defines sight as the most important of the senses, primarily because it allows individuals to share an intersubjective experience and thus creates common social world (Schütz 1964), sensual iconoclasm questions the credibility of sight and concentrates on other senses.

Sociologists researching sensual iconoclasm may be interested in Tom Tykwer's film *Perfume: The Story of a Murderer* (2006). I am referring to the movie and not to the original book by Patrick Süskind, as the movie perfectly exposes the relations between sight and other senses. *Perfume* is a story of an eighteenth-century perfume apprentice, John Baptiste Grenouille. Since birth, Grenouille had a unique gift of a very detailed and accurate detection of scents. As the narrator describes him, "If his name has been forgotten today, it is for the sole reason that his entire ambition was restricted to a domain that leaves no trace in history: To the fleeting realm of scent." This sentence well characterizes the status of the sense of smell—but probably also other non-visual senses, like touch—in social practices. Other senses are marginalized because people do not have abilities to preserve their sensory impressions. If we see something, we can preserve the view on the picture. But we cannot do the same when we smell or touch something because technologies that would allow it are much less common than, for example, photography. That is why these senses "leave no trace in history." Consequently, history lessons in schools usually focus on easily observable events in public sphere and thus emphasize the importance of the sight.

It is worthwhile to pay attention to how the director of *Perfume* tried to communicate the smell by giving the viewers visual impression of scents. When we see a fish market in Paris, the narrator says, "Naturally, the stench was foulest in Paris, for Paris was the largest city in Europe. And nowhere in Paris was that stench more profoundly repugnant than in the city's fish market." The stench is communicated to the audience through different procedures: rapid changes in frames, contrasting colors, unpleasant views (blood, guts, vomit), short and sharp sounds, whereas the images of elegant and full of nice scents Paris district show bright colors and elegantly dressed people. Nice (or at least neutral) sounds and quiet music in the background accompany the consumption of different foods, which are shown in close-ups. In both these examples, the camera movements, together with the visual and auditory experiences, strive to place the viewers inside the movie so they can also share the sense of smell.

These examples relate to the translation from the olfactory messages into visual messages. Filmmakers, however, are aware that such a translation may be questionable—not only in a movie, but also in real life. In another scene the attention is paid not only to the fact that the ocularcentrism can be replaced with the primacy of other senses, but also to the problems of translating from olfactory experiences into other sensory experiences such as sight or hearing sensations. Teenage Grenouille is shown lying on the ground near a water reservoir, while the narrator says, "When Jean-Baptiste did finally learn to speak, he soon found that everyday language proved inadequate for all the olfactory experiences accumulating within himself." Then the camera moves over to show other objects that Grenouille smells (but which he does not see). He tries to identify them by words: "Wood . . . Warm wood . . . Grass . . . Wet grass . . . Stones . . . Warm stones . . . Water . . . Cold water . . . Frog . . . Wet stones . . . Big, wet frog stones . . ." When he is finally smelling frog spawn (which the viewers can see on the screen), he says, clearly upset because of the inability to find adequate words, "Something . . . Something . . . Something . . ." The abstract nature of language prevents Grenouille from putting everything that he receives through the sense of smell into words. However, the director manages to do this, at least partially, when the sense of smell is also referred to the pictures.

Perfume allows a better understanding of sensual iconoclasm, which I defined as questioning the credibility or usability of sight, and thus as a form of opposition against ocularcentrism. First, it suggests that visual experiences are possible to be shared by translating them into texts and create a platform for an intersubjective experience. Second, it points to the difficulty of translating some sensory experiences into other ones.[15] Contrary to sight and the pictures it produces, which seek to create the *commonality* of experiences, we can assume that the purpose of sensual iconoclasm practices—and not only in a cinematographic context—is to *individualize* the human actions. Rafał Drozdowski (2006:194–214) provides examples of such practices when he refers to audiophilia, relaxation massages, or aromatization processes as activities that represent a breach in otherwise ocularcentric nature of contemporary society.

CONCLUSIONS

Sociological discussion on iconoclasm can lead to at least two conclusions. The first one concerns the contemporary iconoclasm, whereas the other applies to almost all iconoclastic practices, regardless of their time period. Therefore, we may assume that on one hand iconoclasm has some unchangeable features, whereas on the other hand it may vary historically and culturally.

The first main conclusion follows Mitchell's (2005:12–13) statement that the destruction of pictures results in the creation of new ones. This often occurs particularly when a picture is a symbol of domination over a place or territory. Destruction of such monuments when related to the transition of a political system is resulted from a desire to symbolically show the change of government and its values. However, in places where old monuments once stood, new ones quickly appeared. The conquest of Berlin in 1945, for example, immediately resulted in replacing the Nazi flag waving over the Reichstag with its Soviet counterpart. The significance of this fact is emphasized by the development of digital technology. September 11 and the collapse of the Twin Towers in New York is visually documented from many vantage points, thus the visuality of the Trade Center's buildings was replaced by the images of their destruction. Similar examples are numerous.[16]

The second conclusion seems to be even more interesting because of its paradoxical nature: iconoclasm, with the exception of sensual iconoclasm, cannot exist without visuality and is based on the faith in the power of pictures and sight. Iconoclasts do not destroy pictures because of their ugliness. They do it because they are convinced that visuality influences other people's activities. As I mentioned earlier, the development of digital media makes the production and distribution of pictures easier and thus increases the visual impact on everyday life. Iconoclasts destroy pictures to protect those who do not have the ability to acquire visual literacy, and are therefore more susceptible to visual persuasion. Therefore, as Jean Baudrillard notices in his book *Simulacra and Simulation* (1994[1981]), iconoclasts are real defenders of visuality.

Why, then, is iconoclasm important from the sociological point of view? The reason is that the study of this phenomenon concerns not only attitudes and opinions about visual sphere, but also behaviors related to it. Of course, it is possible to study them in different ways, for example, by researching the dissemination of pictures on the Internet, creation, and usage of photo albums, or social functions of photography. But the studies of iconoclasm constitute "added value" to the research of visual sphere. First, and foremost, they allow to take into account historical, theological, and legal data and thus to compare the functioning of current visual sphere to past visual spheres. Second, it allows to research also those individuals and groups who do not participate in the production of pictures. Third, it gives an opportunity to combine visual and material culture research. In summary, the studies of iconoclasm allow to research the meanings of pictures and their social functions or even explore the social aspects of the sense of sight as well as other senses. Therefore, they cover all the understandings of the visual sphere presented in the introduction of this paper.

NOTES

1. The paper is a shortened version of one of the chapters of my doctoral dissertation "Social Visual Competence as a Research Problem of Contemporary Sociology" (*Wizualna kompetencja społeczna jako przedmiot badań współczesnej socjologii*). This research project was financed by the Polish Ministry of Science and Higher Education (2009–2010).
2. Following the proposal of Emmison and Smith (2000), in this chapter I understand pictures as any object experienced by sight. Hence, there are both two-dimensional pictures (e.g., photographs, paintings, and maps) and three-dimensional pictures (e.g., sculptures, monuments, buildings, and bodies). However, they may be treated as visual objects only when appropriate actions are directed at them. For example, a building becomes a visual object when it is aestheticized, that is, classified as "beautiful" or "ugly"; a photograph becomes visual when we look at it, but not when we use it as a coaster. This proposal refers to the idea of visual anti-essentialism (Bal 2003) and underlines both socio-cultural and situational aspects of visuality.
3. King James Bible.
4. As this chapter is not a complete monograph of historical and theological aspects of iconoclasm, I treat them as a prelude to further discussion. But it is worthwhile to mention that different aspects of iconoclasm are present also in other religions. For example, in the Koran there is no direct prohibition of producing pictures, but it appears in the Hadiths—the stories from the life of the Prophet Muhammad. According to one Hadith, any artist who created pictures of living creatures will be asked to make them alive during the doomsday. If he or she is not able to do so, he or she will be condemned to eternal ridicule which is a form of hell. Despite the lack of direct prohibition, Islam is opposed to the production of the pictures of God, because that

would be contrary to the basis of Islamic theology, which is the doctrine of unity (Boespflug 2006:39–64).
5. Felix Dzerzhinsky (1877–1926) was the director of Cheka—Bolshevik secret police.
6. Enver Hoxha (1908–1985) was the leader of communist party in Albania.
7. Transformation iconoclasm is related not only to the events from the second half of the twentieth century, but also to later periods. The seizure of Baghdad by the United States troops in 2003, during the second war in Iraq, was also symbolically confirmed by the destruction of the statue of Saddam Hussein. However, it is worth mentioning that the crowd in Baghdad was probably not spontaneous, but intentionally collected by U.S. troops. One can even say that the pictures of executions played a similar role. Pictures of the body of the Romania's former president Nicolae Ceausescu, or a video showing the hanging of Saddam Hussein, could also be defined as the destruction of a certain symbolic picture of the regime.
8. http://niewidzialnemiasto.pl (retrieved July 5, 2011).
9. http://www.indymedia.org (retrieved October 28, 2011).
10. http://www.youtube.com/watch?v=MzaN2x8qXcM (retrieved March 30, 2012).
11. http://whitehouse.georgewbush.org/index.asp (retrieved July 5, 2011).
12. http://www.whitehouse.gov/ (retrieved July 5, 2011).
13. http://www.fuckhotmail.com/ (retrieved July 5, 2011).
14. Then it was http://hotmail.com, now it has been changed to https://login.live.com/ (retrieved July 5, 2011).
15. Both of these facts may explain why it has been visual sociology (and not, e.g., olfactory sociology) that has been recently developed within social science. However, methodologies which are based on other (non-visual) senses are becoming more and more popular in social science (Pink 2009).
16. The results of iconoclastic practices are spectacular; they are willingly viewed by a mass audience and are even desirable. A good example can be the popularity of disaster movies. Thus, this can be just a kind of voyeurism, desire to watch something extraordinary and sometimes forbidden.

REFERENCES

Bal, Mieke. 2003. "Visual essentialism and the object of visual culture." *Journal of Visual Culture* 2(1):5–32.
Baudrillard, Jean. 1994. *Simulacra and simulation*. Ann Arbor: University of Michigan Press.
Boespflug, François. 2006. *Caricaturer Dieu? Pouvoiras et dangers de l'image*. Montrouge: Bayard.
Bourdieu, Pierre. 1984[1979]. *Distinction: a social critique of the judgment of taste*. London: Routledge.
Bourdieu, Pierre. 1990[1965]. *Photography: a middle-brow art*. Stanford, CA: Stanford University Press.
Bryer, Anthony and Judith Herrin, eds. 1977. *Iconoclasm: papers given at the Ninth Spring Symposium Of Byzantine Studies*. Birmingham: Centre for Byzantine Studies, University of Birmingham.
Burgin, Victor. 1996. *In/different spaces: place and memory in visual culture*. Berkeley: University of California Press.
Carpo, Mario. 2001[1998]. *Architecture in the age of printing: orality, writing, typography, and printed images in the history of architectural theory*. Cambridge, MA: MIT Press.

Certeau, Michel de. 1984. *The practice of everyday life.* Berkeley: University of California Press.
Chase, John, John Kaliski, and Margaret Crawford, eds. 2008. *Everyday urbanism.* New York: Monacelli Press.
Crary, Jonathan. 1999. *Suspensions of perception: attention, spectacle, and modern culture.* Cambridge, MA: MIT Press.
Debord, Guy. 1994[1967]. *The society of the spectacle.* New York: Zone Books.
Drozdowski, Rafał. 2006. *Obraza na obrazy. Strategie społecznego oporu wobec obrazów dominujących.* Poznań: Wydawnictwo Naukowe UAM.
Edwards, Elizabeth and Janice Hart, eds. 2004. *Photographs, objects, histories: on the materiality of images.* London: Routledge.
Emmison, Michael and Philip Smith. 2000. *Researching the visual. Images, Objects, Contexts and Interactions in Social and Cultural Inquiry.* London: Sage.
Forrester, Michael. 2000. *Psychology of the image.* London: Routledge.
Freedberg, David. 1989. *The power of images: studies in the history and theory of response.* Chicago: The University of Chicago Press.
Gamboni, Dario. 1997. *The destruction of art: iconoclasm and vandalism since the French Revolution.* London: Reaktion Books.
Garland-Thomson, Rosemarie. 2009. *Staring: how we look.* New York: Oxford University Press.
Giddens, Anthony. 1991. *Modernity and self-identity: self and society in the late modern age.* Stanford, CA: Stanford University Press.
Goffman, Erving. 1963. *Behaviour in public places: notes on the social organization of gatherings.* New York: The Free Press.
Goffman, Erving. 1971. *Relations in public: microstudies of the public order.* Ringwood, Victoria: Penguin Books.
Hall, Stuart. ed. 2000. *Representation: cultural representations and signifying practices.* Glasgow: The Open University.
Hirsch, Marianne. 1997. *Family frames: photography, narrative and postmemory.* Cambridge, MA: Harvard University Press.
Jacobsen, Michael H. 2009. *Encountering the everyday: an introduction to the sociologies of the unnoticed.* Basingstoke, UK: Palgrave Macmillan.
Jay, Martin. 1998. "Scopic regimes of modernity." Pp. 66–69 in *The visual culture reader,* edited by N. Mirzoeff. London: Routledge.
Jenkins, Henry. 2006. *Convergence culture: where old and new media collide.* New York: New York University Press.
Lalvani, Suren. 1996. *Photography, vision, and the production of modern bodies.* Albany: State University of New York Press.
Latour, Bruno. 2002. "What is iconoclash? Or is there a world beyond the image wars?" Pp. 16–38 in *Iconoclash: beyond the image wars in science, religion and art,* edited by B. Latour and P. Weibel. Cambridge, MA: MIT Press.
Maffesoli, Michel. 1996. *Time of the tribes: the decline of individualism in mass society.* London: Sage.
Major, Patrick. 2010. *Behind the Berlin Wall: East Germany and the frontiers of power.* Oxford: Oxford University Press.
Manovich, Lev. 2001. *The language of new media.* Cambridge, MA: MIT Press.
Messaris, Paul. 1994. *Visual literacy: image, mind, and reality.* Boulder, CO: Westview.
Mitchell, William J. T. 1984. "What is an image?" *New Literary History* 15(3):503–537.
Mitchell, William J. T. 2005. *What do pictures want? The lives and loves of images.* Chicago: The University of Chicago Press.
Pink, Sarah. 2009. *Doing sensory ethnography.* London: Sage.

Rose, Gillian. 2001. *Visual methodologies: an introduction to the interpretation of visual materials*. London: Sage.
Schütz, Alfred. 1964. *Collected papers*. Vol. 2. *Studies in social theory*. Edited by Arvid Brodersen. The Hague: Martinus Nijhof.
Turner, Victor and Edith L. B. Turner. 1978. *Image and pilgrimage in Christian culture*. New York: Columbia University Press.
Tyner, Kathleen. 1998. *Literacy in a digital world: teaching and learning in the age of information*. Mahwah, NJ: Lawrence Erlbaum Associates.
van Asselt, Willem, Paul van Geest, Daniela Müller, and Theo Salemink. 2007. "Introduction." Pp. 1–29 in *Iconoclasm and iconoclash: struggle for religious identity*, edited by W. van Asselt, P. van Geest, D. Müller and T. Salemink. Leiden: Koninklijke Brill NV.

5 Picturing "Gender"
Iconic Figuration, Popularization, and the Contestation of a Key Discourse in the New Europe

Anna Schober

Over about the last forty years philosophical and sociological concepts of gender have developed and consequently undergone transformations. Although in most cases "gender" is used as a deconstructive philosophical or sociological concept that tries to reveal that every identity is the result of a historically contingent cultural construction, its definition remained continuously open and was subject to various disputes, discussions, disavowals, and redefinitions and struggles. Thus in the 1970s (and in some academic and political milieus longer and with varying accentuations), for instance, gender as a "cultural construction" was often contrasted with "sex"—which was then used as indicating the biological levels of our lives as sexual beings. Gayle Rubin (1975), for instance, investigated a "sex-gender" system in which the two levels refer to each other but are not causally or functionally related. In the 1980s (Jaggar 1983:109) and later most prominently in Judith Butler's *Gender Trouble* (1990) such definitions were, however, challenged by demonstrations which revealed that what we usually see as biological and natural is also culturally determined.[1]

Linked to such discussions in the academic field, gender soon—but in an always milieu-specific way and marked by strong varieties and sometimes even by contestations and struggles—started to enter the realm of the political and to replace previous notions used in political activism such as "women" or "feminism" (Hark 2005:37). In the last two decades gender has had a particular transnational success—it has become a central issue for EU and worldwide social reform programs—for instance, under the buzzword "gender mainstreaming"—but also for political grassroots movements. This goes along with a popularization of gender discourses—a popularization that occurred in a world increasingly made of images and one that depends heavily on quick, captivating communication via visual creation. This has led to the translation of concepts of gender into visual worlds such as public political parades, festivals, exhibitions, fanzines, or websites. Parallel to this, gender has also stimulated adoptions in the visual and performance arts and in video and film production.

In this chapter[2] I investigate a variety of examples of popular and artistic adoptions of concepts of gender in diverse European cultural contexts.

In my analysis of the public life of gender I try not to separate the realms of public space and culture, as is often the case, but to show their mutual interlacement. In doing so, I demonstrate that political initiatives depend on cultural and also artistic practice in order to address their target audience and that art or cultural actions are often inspired by scientific and political concepts and ideas.[3]

At the same time I focus on the double-edged role that images perform in the course of generating a public life for gender: on the one hand, they increase the public presence of gender as a reflexive theory, but on the other, this figuration of gender through visual culture holds another—somehow antidromic—tendency. Visual worlds also have the capacity of putting political, philosophical, and sociological concepts and definitions into crisis.

A REARRANGEMENT OF THE SCENERY

Since the late 1980s, a "rearrangement of the scenery" can be observed in the ambit of the political creation of European society (Hark 2005:37). An emancipatory discourse promoting women has been replaced by one that features gender. This notion has—as was formulated in 1995 by the Contact Group on Gender during the United Nations Meeting on the Status of Women in Beijing—"evolved as differentiated from the word 'sex' to express the reality that women's and men's roles and status are socially constructed and subject to change" (quoted in Butler 2004:182). But even if such concepts were quite quickly adopted on a transnational level by a variety of social agents as a way of addressing the social construction of women and men—especially in the U.S., Germany, and parts of northern Europe—this has remained to the present an intensely debated subject in scientific as well as in political and religious discourse and is often contested. At the UN Meeting in Beijing, for instance, the Vatican as well as queer groups voted for the use of sex instead of gender (Butler 2004:183). And also within the European Union the adoption or rejection of gender was debated on various fronts (Stratigaki 2005). In particular, the fact that concepts of gender as well as the bulk of scholarship accompanying them originate mainly in the Anglo-Saxon world and have been somehow "imported" from the U.S. context led to clashes with many European feminist cultures (Braidotti 2002). Nevertheless, such concepts were further disseminated with the proclamation of gender mainstreaming as an official goal of the equal opportunities politics of the European Union in the Treaty of Amsterdam in 1997 (Shaw 2002)—which enhanced the need to "translate" what is meant by gender into attractive figures in order to bridge the gap between the scientific-political discourse and the wider public. Some local traditions of dealing with women's emancipation have become marginalized by the spread of this concept—such as

Italian feminist groups that strengthened the concept of *differenza sessuale* ("sexual difference") and which have in the 1990s increasingly closed themselves off from diverging feminist discussions (Putino 1998). Other and especially younger generations of political grassroots movements, however, have embraced culture-oriented interpretative models such as gender and sex-gender.[4] Especially political groups of "third-wave feminism"[5] such as the "Riot Girrl movement" and the "lady-fests" showed themselves as being inspired by gender and translated these concepts into music, fanzines, film and multimedia festivals, or spectacular party events. And in the 1990s artists stimulated by culture-oriented, constructivist interpretative models also began to invent unusual figurations in relation to the body and our life as sexual beings or to relate to other art that evolved around such notions. In parallel, queer or transgender groups started to oppose gender and to create self-representations featuring sex.

In the following section I focus on how such a "rearrangement of the scenery" from a discourse representing women to one featuring gender can be traced in the realm of public visual culture. This reorganization is never complete, nor does it proceed in a homogenous way. On the contrary, usually various political positions, different artistic and media initiatives, or official reform strategies in respect to parity or issues such as sexuality, the family, and work conditions simultaneously compete in attracting an audience. What changes is the hegemonic position these initiatives may gain—whereby any hegemonic victory of certain meanings of gender always remains precarious and can become subject to further questioning.[6]

IMAGES AND THE TRANSITION FROM ONE REGIME OF CERTAINTY TO ANOTHER ONE

Our relationship toward our own sex or toward that of others is one of certainty, conviction, and passionate practice (Zerilli 2005:38). This relationship has less to do with knowledge, in the sense that we are searching for proof to justify our actions as sexual beings and thus cannot be easily influenced via propositional statements. Today we know that there are no ultimate criteria for differentiating sex. For example, in determining competitors' sex the Olympic committee changed from a naked inspection to chromosome tests and back to a naked inspection. However, in most cases we usually have no doubt with regard to the sex of a counterpart or of our own sexual identity. Certainty, according to Wittgenstein, is a doing not a knowing (Wittgenstein 1984:127). We inhabit a house in which the existence of two sexes is backed by various details of daily life practice. We are guided by certainty and passion without analyzing every move we make or constantly questioning our actions. Hence the general framework of the game we are playing in this house is certainty not knowledge.

In recent decades an enormous production of knowledge with regard to masculinity and femininity has been counter-posed to this matter-of-factness and the connected myths. It sought to demonstrate that gender is a becoming, a collective, constantly repeated, and thus transformed construction.[7] However, the question is, how can this knowledge affect the certainty that guides our everyday actions and the passions and drives connected to them? Even if today we are increasingly confronted with strategies and tactics that relate to us as a sexual being (as part of social and political reform, health care, work policies, advertisement, sports, the arts etc.), we still continue to live the lives we find ourselves thrown into as men and women with all their turbulences. To detach oneself from the certainty and the matter-of-factness that guides our practices in everyday life and to assume other ones constitutes a conversion—is comparable to abandoning religious or ideological beliefs and assuming others. This process usually involves being profoundly touched, set out of oneself, fascinated and involved, rather than resorting to critique, explanation, and demonstration. Hence there is also a gap between the certainty of the everyday and the knowledge of gender studies, even if passionate involvement and rational judgment are not mutually exclusive, as some traditions of political science would make us believe,[8] but rather reciprocally inspire and motivate each other.

This gap between the certainty of everyday life and knowledge also has to do with the fact that myths cannot be conquered by argument, for instance, by exposing them as "groundless illusions" or by showing that what they tell is "wrong." Myths are composed of a multitude of mutually reinforcing narratives and a variety of often very seductive details that have the capacity to accommodate our wishes, fears, or desires. These narrations exist and circulate without having to be justified and are able to explain and give sense to most of our actions. This also means that critique cannot be formulated from a position that lies completely "outside" mythical narratives but only from one that remains involved and tries to give an approximate account of transformations in our ideas, scientific concepts, ideologies, and myths and of the distinctions that can be made between them. However, as part of such processes, as well as of the changes of what is able to accommodate our desires and fears, myths do indeed transform themselves: we are involved in a more or less constant innovation of the mythical (Schober 2001:40, 158), with some myths that no longer correspond to what we experience (or what we desire or fear) losing their importance and even being abandoned and others being revised, fused, or even revitalized or reinvented. The question I propose to address here concerns which innovations in our mythical figurations are related to gender and how this is tied to processes of figuration and image acts.

In turning our attention toward these processes of innovation and conversion from one mythical world to another, a particular faculty comes to the fore that in a very eminent sense is responsible for them: the faculty of imagination, as Cornelius Castoriadis (1997) calls it. According to Castoriadis, this faculty allows us to generate forms and figurations that are not already given by experience or the existing order of things. Such "innovative" figurations are capable of fascinating, involving, and affecting us as viewers and can in this way transform institutional society through captivation and conviction.

This role of images in setting existing convictions, norms, and everyday certainties into crisis and causing transformations in the realm of imagination as well as of social practices was recently brought into the foreground and analyzed by the new discipline of "visual culture studies" (Didi-Huberman 1992; Mitchell 2005; Marin 2007; Mersch 2007). Louis Marin (2007), whose writings are a central reference point for this kind of research, has directed our attention in particular toward the founding, promotional, activating, challenging, and authenticating force images embody by combining a productive agency with a representational effect. This force, for instance, becomes graspable through the authority that some details of images can incorporate for us in the course of the viewing process. Besides, images also exert power in providing fantasy images in which we as subjects are able to view and examine ourselves and in which our desire returns as that of a (repeatedly) other self. Finally, the force of images can be measured by paying attention to events of transition, by tracing reception histories and processes of adoption, and by studying where they become issues or even sites of struggle.

These methodological reflections provided by visual culture studies show that images are not a secondhand representation of something primary but a generating force. At the same time they again point out that images cannot only be seen as active forces that work to strengthen social norms but can also become agents causing a crisis of norms and political concepts.

In contrast to this, dominant strands of gender studies that started to spread in the 1990s and are following Judith Butler's *Gender Trouble* (1990) are characterized by the fact that they are usually engaged to a large extent with norms in relation to the body and its social performance. Often, these approaches of gender studies also focus on images as active agents in enforcing social norms, including gender norms.[9] The second potential of images of setting existing norms and concepts into crisis was largely ignored. A further effect of this is that popular (and artistic) image making is often discussed in gender studies or cultural studies in binary models: either they are seen as maintaining or enforcing gender norms or they are judged as "subversive" or even "critical."[10] The dissolution of norms that characterize "late" or "post-modernity" (Hegener 2009:132) as well as innovations in mythical explanations and events of conversion between belief-systems recede into the background.

In this text my approach aims at integrating visual culture studies for the analysis of the public life of gender. The investigation focuses not on image worlds as products but as acts that trigger processes of reception or become sites of struggles, and on how pictorial worlds are linked to those passages in the mythical and ideological realm by which people make sense of their lives. Furthermore, I analyze how such passages, in turn, are reflected, negotiated, and contested by the production of new and pronouncedly "different" images concerning masculinity, femininity, androgyny, sexuality, the family, and genealogy. I thereby counteract the usual practice of fixing the meaning of gender. In contrast, I will focus on how concepts of gender have been adopted for the creation of a huge variety of image productions within the various European cultures through which they have passed in recent decades.

In doing so, the following conceptual definitions shall apply in this paper. On the one hand, gender is seen as a very specific concept that in itself already contains a set of preliminary considerations: it opts for a societal analysis; it reanimates the dichotomies nature/culture or biology/society; it renounces the representation of sexual relation and its conflicts in favor of a voluntary abstraction; it constitutes a subject of thought that seems to have a certain autonomy (Fraisse 1996:53). But on the other hand, gender is perceived in its function as a "figure of the newly thinkable" in terms set out by Cornelius Castoriadis (1997). As such, gender concepts animate the political imagination, which leads to collective bricolage and to vivid processes of iconic figuration. The images and visual worlds created in this way disseminate and popularize this concept further.

Gender or sex-gender are in this way seen as particular figures of thought able to trigger imagination and action, i.e. they are situated beyond "the epistemic demand of deciding the true and the false" (Zerilli 2005:59).[11] On an epistemic level, rather than using the sex-gender distinction, I use the concept of "sexual difference," which however does not propose an absolute difference but operates on the assumption that men and women (as well as transgender persons or exponents of other identity groups) are different and equal at the same time—something that needs no further definition because any definition will set new social norms (Fraisse 1996:135f.). At the same time it is exactly this "and" instead of an "either/or" that poses the biggest challenge for the political and public handling of our life as sexual beings.

THE DISSEMINATION AND AGGLOMERATIONS OF IMAGES

Gender usually gains public presence via the medium of the image. However, iconic figurations of such culture-oriented interpretative models are usually characterized by a particular tension: On the one hand, there is no ready-made iconic tradition available for the act of figuration to link itself up to, that is, an iconic tradition in relation to these new concepts is "invented" only in the course of the dissemination of these concepts. But on the other hand, the various acts of iconic figuration indeed refer back to existing iconographic motives, visual conventions, myths, and narrations, which are in this

Picturing "Gender" 63

way adopted, recast, and altered. The most important of these existing iconic motives that the various acts of pictorial figuration tend to link up to are the "new man" and "new woman," the androgynous, the hermaphrodite, or other sexually liminal and intermediate creatures, and the machine-body and visions of the self-sufficient and self-(re)creating body.

A first overview of iconic figurations that the concepts of gender inspire brings to light that several examples show in one or the other point a similarity and closeness vis-à-vis each other. In this way it is possible to identify the dominant "agglomerations" or "clusters" of bodily image-creations in relation to culture-oriented interpretative models such as gender. But even if such agglomerations or clusters can be identified, they never exist neatly separated from each other but overlap, that is, several images take part in more than one of them. In the following, I will present a range of such image clusters that have emerged in relation to gender since the 1990s. I do however not subscribe to a notion of something like a "completeness" of documentation but rather aim at giving a comprehensive overview and at depicting exemplary case studies.

Pictograms and the Assertion of a Neutrality of Reform

Quite often posters, brochures, folders, or websites created as part of public campaigns in relation to gender issues show icons and pictograms. Often we see the now quite standardized symbol for man and woman: a circle with either a cross or an arrow attached. Other common pictograms that play with male and female are, for instance, small manikins similar to those we find on the toilet doors of bars or restaurants. With concepts of gender, usually icons figuring man and woman appear together, even if sometimes alterations are invented, for example, a circle with both arrow and cross or manikins that show re-combinations or the crossover between the icons for man and woman. Often, colors are used in relation to these icons and pictograms that are traditionally, that is, mainly since the 1970s, related to feminist public action: purple and pink.

These figurations demonstrate the rootedness of gender discourse in a tradition of reform discourses and modernist political movements—because in these contexts pictograms were first used to address the masses. In the Soviet Union, after the revolution, communist reform organizations used icons to overcome language barriers and illiteracy and to direct "the masses" into specific directions (Zimmermann 2006:43). In Vienna in the 1920s, Otto Neurath of the Vienna Circle invented the "Isotype," a system of schematic, simplified pictures in order to manage social life in an internationally unified way (Lupton 1989:148). And since the 1970s, the feminist movement used graffiti adoptions of such icons in order to create public presence for solidarity between women.

The figuration of gender in the form of pictograms reiterates such a tradition of a "universal" picture language which emerged most strongly in the period of "heroic modernism" (between the First and the Second World War) and is the product of a modernist belief in the reduction of information, the simplicity of form, and standardization for the sake of saving time, labor, and money, and of communicating effectively with "all" members of society (Kinroos 1989:138). It is thus driven by a commitment to the rational, the progressive and the economies of effort. Because central features of this modernist belief system, such as ideas of a rationality of public processes or the effective possibility to reform citizens "from above," have been challenged in recent decades, this tradition is now performed in a transformed way. Contemporary designers often combine pictograms with more emphatic, emotionally appealing image components. In one example[12] we see the icons for man and woman (a circle with a cross and another with an arrow) in the form of wooden sculptures set in a flat landscape with a wide horizon under a blue sky and small friendly white clouds—which suggests a harmonious co-existence of the male and the female as part of an "optimistic" landscape.

Beyond what is explicitly represented however, these pictograms usually continue a modernist tradition that asserts and promotes a "neutrality" that often goes together with a visualization of symmetry and equivalency or "parity." Such an assertion, promotion, and creation of visual evidence of neutrality in respect to the relations between the sexes and their public management pushes into the background any difficulties, conflicts, or problems that exist in these relations or that may come up in public reform strategies.

Symmetry

Concepts of gender are often translated into images that show a pronounced symmetry.[13] This is demonstrated in a photo that first appeared in a web article headlined "Gender Equality in Sweden" in 2008 and was then adopted by a number of other websites and blogs.[14] It shows the close-up, sharp profiles of a male and a female face turned toward each other against a blurred background. This sharp profile and the symmetrical arrangement of eyes, mouths, and nostrils makes the faces seem strangely flat, like the mirrored wings of a butterfly. In this way, the faces seem similar, androgynous, of equal value. By contrast, the extreme close-up invites us to investigate capillary differences in this symmetry, differences which are nevertheless constantly pushed back by the symmetrical pattern of the whole. Furthermore, the two faces gaze into each other's eyes, which despite all similarity, symmetry, and relatedness also creates a certain dynamic tension. Hence the image stages a symmetrical and at the same time dynamic co-presence of the male and female, where both parties are of equal value—it promotes such a co-presence and provides it with authority, which could be the reason for the multiple adoptions triggered by this image. Also here, this assertion and promotion of symmetry and parity makes conflicts and hierarchies between the sexes disappear from the public stage.

At first glance the photo series EINE GLÜCKLICHE EHE (A HAPPY MARRIAGE, work in progress: 2003–2011) by the Italian-German artist Daniela Comani fits into this analysis.

The series consists of images of a man and a woman of the same height, same figure, arranged symmetrically in front of the camera: sometimes standing and holding hands, sometimes side by side in bed, reading. Their clothing also seems interchangeable, modern, and casual: unisex clothing loosely oriented on masculinity—jeans, slacks, shirts, T-shirts, and sweatshirts. A second, closer look is somehow unsettling: the two people are too similar, their clothing too interchangeable, and the mirroring of one in the other appears overstated. Finally, we realize (through the catalogue, Internet, or intuition) it is always just one person, the artist herself alternating between woman and man by re-sampling the same everyday outfit and slightly shifting her posture or legs, or by wearing a false three-day beard. Afterwards, all these self-portraits are digitally compiled—with the medium of photography being complicit in outsmarting us as an audience because it is associated with authenticity and the representation of something antecedent.

Figure 5.1 Daniela Comani, from the series "EINE GLÜCKLICHE EHE" (A HAPPY MARRIAGE): #2, 2003, © Daniela Comani.

This work by Daniela Comani highlights a not yet mentioned aspect of images presenting a symmetrical co-presence of male and female characters and exposes a dimension of our public dealing with parity and difference that often goes unnoticed. As a beholder of these images, I am able to compare myself with the female and male part of the couple represented; I can identify or switch between them. Outwardly there is almost no difference between the two figures. The married couple is transformed into a "unisex couple" consisting of a single person happily married to herself but appearing alternately as a man and a woman. Thus we have a reactivation of myths and desires that refer to the combination of the sexes in one subject—something Roland Barthes (2008:186) called the "myth (and utopia) of the androgyne." In this photo series this myth is not represented as a spectacular but as a discrete public appearance, that is, difference, desire, repulsion, the experiences of delineation and belonging, and the ambivalences and conflicts connected with them take place within the characters—something with which we as an audience are invited to empathize.

Thus Daniela Comani's photos indicate a central aspect of how we inhabit public space, for, as Zygmunt Bauman (1991:96) has shown, our public life has in late modernity become increasingly characterized by interiorizing and privatizing difference, the ambivalences of belonging, and the conflicts connected to this. A HAPPY MARRIAGE situates itself as part of this tendency. This work presents new "neutral" or "androgynous" bodies characterized by a dissolution, blurring, and rearranging of the exterior signatures of the male and female and a privatization of differences, ambivalences, and conflicts. Hence in this example, too, a pronounced staging of symmetry is accompanied by an absence of conflicts and hierarchies—this time they vanish into the interior worlds of the represented "unisex" persons.

"Having Everything"

Another agglomeration of images invented in more or less explicit connection with gender emerges around the topic of the hermaphroditic body, that is, these image-worlds usually expose successful combinations of fragments of female and male body parts. An example of this is *Passage* (2004–2005) by the British-Scottish artist Jenny Saville.

Like some of her other paintings it is larger-than-life and shows a disquieting, fleshy body of jarring proportions. Saville herself has been interviewed often about her way of presenting (mostly female) nudes. In relation to the way she uses the notion of gender in *Passage* she says,

> I was searching for a body that was between genders . . . I had explored that idea . . . of floating gender that is not fixed . . . Thirty or forty years ago this body couldn't have existed and I was looking for a kind of contemporary architecture of the body. I wanted to paint a visual passage through gender—a sort of gender landscape. To scale from the

Figure 5.2 Jenny Saville, "Passage" (2005), © Jenny Saville.

penis, across a stomach to the breast, and finally the head. (interview with Schama 2005:126)

Saville's way of painting explores the tactile pleasure of paint application, that is, deals with "paint as substance" (Damisch 1979:12). She creates sensory arrangements of blues, whites, and beiges that function simultaneously as solid markers building the body up. In this way, Saville's paintings invite us as viewers to embark on a self-examination of bodily sensations in relation to our sexual being and to explore societal conventions and taboos.

The enormous scale enhances the impression of being involved in a physical space, that is, in a lived-in, fleshy body that emits sensations. In this way *Passage* provides its viewers with a sensation of "having everything" or "being both," which means being male and female at the same time. The excitement of having everything is expressed explicitly by the artist: "penis and breasts all at the same time. It's electric, it's like wow! To see something in a way you have not looked at it before" (Saville in Mackenzie 2005).

In this painting Saville does not mainly depict the reality of a specific (e.g., transgender or intersexual)[15] body but rather creates and authenticates a new kind of body, a "contemporary architecture of the body" as she calls it. A central feature of this contemporary architecture of the body seems to be the desire to "have everything," to be male and female at the same time—and this in a proudly presented way. Paintings such as *Passage* not only enable the beholder to switch between male and female identifications but allow both female and male viewers to try to experience a kind of bodily sensibility of what it would be like to have both breasts and a penis. We see and somehow experience a kind of "collection" of male and female sexual codes in a single and "whole" body—these codes mutually enforce and stimulate each other rather than cancel each other out or propose a "deconstructive" interpretation. The body appears as hypersexual and able to satisfy multiple desires (male and female, heterosexual, transsexual, and gay, for affirmation as well as for critique).[16]

This painting again relates to myths of the androgynous—for example, those that stage the unification of two opposing principles in one being that in this way becomes more "perfect" and "whole" than usual humans.[17] But *Passage* also introduces an alteration: it refers us back to the body and its sexual organs, that is, male and female do not appear so much as "united" but "co-present" in a body "offering" itself to the gaze of the beholders—something that is indicated by the subject's reclined and open position. Hence the painting accentuates the sensuous and erotic aspect of "having both" and presents this—again faithful to this tradition—as signatures of a bodily being that is proudly and perhaps also "superiorly" staging itself.

The "perfection" of being both male and female at the same time is also represented in other examples of contemporary visual production. The acclaimed film *XXY* (2007) by the Argentinean director Lucía Puenzo presents the story of Alex, a teenager born with an intersex body and raised by her/his parents in a kind of drop-out retreat far away from "civilization" on the wild Uruguayan coast. In *XXY* we never see the sexual organs of the protagonist, but encounter substitutes that testify to the "perfection" of what they see: the father, Alex's teenage lover, and a male gang that violently throws down the protagonist and undresses her/him to confirm the rumors with their own eyes.

Visual productions such as *Passage* or *XXY* have a creating and authenticating force. They do not mainly represent hitherto marginalized, real existing transvestite or intersex persons (who usually do not have sexual organs of both sexes in such a "perfect" way) but rather give presence and authority to a new experience of the hypersexual[18] linked to a collecting of sexual sensibilities connected with the male and the female. They offer visual markers that also give authenticity to such an experience.

Passage is a result of the artist's involvement in various contemporary discourses. Saville uses painting to make sense of bodily functions that she observes in her everyday surroundings, medical books, plastic surgery imagery, magazines, the Internet, films, and productions by other artists. She even gets inspiration for her paintings from watching plastic surgeons at work (interview with Schama 2005:124). Thus her paintings pick up and rework various "found" images and bodily postures, whereby they transpose what she witnesses into what Simon Schama calls "the anatomy of paint" (interview with Schama 2005:126). Thus *Passage*—like other paintings by the artist—is able to stimulate the viewer's interpretation, not only by means of her manner of color application but also by interacting with this multiplicity of contemporary visual productions. In doing so, her paintings trigger various stories of involvement. Saville, for instance, gets fan mail from overweight people, which means that there is a "fat-pride discourse" connected with her paintings.[19] At the same time a feminist discourse rejecting and discussing body norms or a transgender-pride discourse is also linked to her works. In addition, an art discourse that features painting in contrast to installation art also asserts itself through her paintings. *Passage* acts as a kind of "bridge" between these discourses, letting viewers move between them by interrogating their sensations or to just remain with one of them and become interested (or irritated) by the others. These various discourses overlap only in the sense that they connect to Saville's paintings and highlight one or the other aspect of them—otherwise they can be quite remote from each other, as the art and transgender-pride discourses are. But *Passage* not only offers points of entrance for these discourses. The painting also has the capacity to throw these discourses, their concepts, and the political efforts connected with them into confusion. This can be seen in discussions of this painting on transgender websites[20] that express a sense of disturbance. Some viewers experience a variety of sometimes very conflicting bodily sensations rather than pride vis-à-vis this artwork. Like other paintings by the artist, *Passage* evokes "gut feelings," ambivalent identifications and even disgust (Meagher 2003) and in this way blocks efforts to set up clear us-and-them communities.

The Patchwork Self in Transformation

Often, images related to gender explore change and transformation in relation to bodily appearance. Paradigmatic for this are the photographs by U.S. artist Cindy Sherman, one of the protagonists most frequently featured in connection with gender. Interestingly it is not she herself who makes this connection but critics who often interpret her work as a distancing restaging of conventional codes of femininity (Mulvey 1996:70). In interviews Sherman either invalidates these interpretations, stating, "they have nothing to do with why I made the work" (Taylor 1985:79), or even rejects them, saying, "I just happen to illustrate some theories" (Helfand 1997). In contrast she highlights the pleasure and fun she experiences in transforming in front of a mirror and how important it is for her to control the final result of the metamorphosis she is creating in this way. The latter is also the reason why she uses herself as a model—because she couldn't control others as she can control herself (Kallfelz 1984:45; Francblin 1992:18).

Both aspects she mentions—the pleasure and fun aspect as well as the desire to control—appear strongly in the vivid reception and adoption history her art has triggered. Several European artists of a younger generation refer explicitly or implicitly to her work. In contrast to Sherman they often closely link a restaging of themselves in new male and female outfits with a critical scientific discourse about gender—such as for example the already mentioned Daniela Comani who in one of the images of the series EINE GLÜCKLICHE EHE (A HAPPY MARRIAGE) appears reading Judith Butler's *Gender Trouble*.

In doing so, several of these artists present the self as a kind of "patchwork self" that can repeatedly and emphatically be reinvented and re-sampled. Usually this reinvention and re-sampling is presented as the outcome of a spectacular, successful, vanguard, and empowering activity that is somehow under the "control" of the doer and is this way purged of all painful, problematic, disordering, confusing, and incontrollable aspects. In a picture of the DJ "Jake The Rapper" featured in the Gender Bender Festival catalogue (2007) we see the bringing together of bodily codes and fragments quite far from each other: high-heeled jackboots in tiger look, a trench coat over a naked, slightly obese body, a tattoo and a classical "antique" pose in the context of a petty bourgeois living room.

Like in a Dada montage, fragments of male and female bodily appearance and their recombinations are picked up as a positive trope or even as one holding the capacity of "revolutionizing" contemporary body appearance.[21] This is also sustained by the textual statements accompanying such body collages. The Gender Bender Festival, for example, explicitly highlights also on the textual level a patchwork identity, translation, transfer, or transformation. In the catalogue of the edition of the festival in 2005 the work "Metamorphose" by U.S. artist Matthew Barney is called a "celebration of the indistinctive and alterable where gender and sex play an important role."

Some of these textual and pictorial figurations explicitly refer to the new technologies and their potential to expand the radius of human action. For instance, the catalogue of the very influential exhibition "Posthuman" (Turin, Athens, Hamburg, 1992–1993), which highlighted forms of self-

Figure 5.3 "Jake The Rapper," Gender Bender Festival catalogue (2007), © Harald Popp and Jake the Rapper.

construction related to (bio-)technologies and the arts, states, "the search for the absolute 'true' self has been replaced by a constant scanning for new alternatives" (Deitch 1993:35). Further,

> many of the most interesting younger artists are dealing with the new conceptions of the body and the new definitions of self that the vanguard of our society is also dealing with. They are exploring through their arts the same questioning of traditional notions of gender, sexuality, and self-identity that is taking place in the world at large. (Deitch 1993:42)

The alterable in this way appears as another central face of "contemporary architectures of the body." Hence when the "Posthuman" catalogue states,

"It is normal to reinvent oneself," (Deitch 1993:19) it also vividly demonstrates that there seems to be a kind of "re-naturalization" or "normalization" of self-transformation happening in connection with a popularization of gender discourses.

These figurations of gender show strong parallels to contemporary consumer and lifestyle culture, which also highlights the potential to reinvent oneself via styles, consumer goods, and new sense-providing belief systems, refurnishes us constantly with "new" sets of life-worlds and ideas for such recreation, and presents this as a successful activity free of problematic aspects. The demands present in most discourses which surround us usually employ notions of "flexibility," "lifelong learning," "throwaway society," "patchwork families," "period of life partnership," or "new nomadism." A celebrating and empowering representation of the self in constant transformation as it appears in most of these image-worlds can thus also be seen as a kind of symbolic "resolution" of problems and crisis in connection with these imperatives. At the same time these image-acts are sites where these new tendencies as well as new potentialities connected with them or particular phenomena such as aesthetic surgery or reproduction technology can be reflected, negotiated, disavowed, or challenged. In this way these figurations also testify to a new openness of contemporary being in relation to the potentialities of life in its sexual dimension and to the expansion of these potentialities via new technologies—which once again highlights the double-edged character of these image acts.

"Kill Your Gender!" Attacking Incorporations of Body Norms

Surprisingly, quite a lot of images invented in connection with gender show Barbie, the famous long-legged, blonde doll. As part of the Austrian Green party's student election campaign in 2009 we see a poster of Barbie in a doll's kitchen with the slogan "break down role models!" In another example, we see her in "symmetrical" combination with Ken, her male counterpart, on a brochure advertising a gender-education meeting for school teachers in Italy. But she also appears in the above-mentioned "Posthuman" catalogue in relation to "artificiality" and a fading away of "reality as we know it" (Deitch 1993:51). In all these visual creations Barbie is quoted not in order to promote her, that is, to present her as a role model to identify with, but as an object that incorporates a norm that today's viewers can interrogate and possibly also free themselves from. This is clearest in the Austrian Green's poster, where further slogans say "this means you" and "betrayed."

In spite of these often very clearly expressed intentions, however, these image-text montages designed to question conventions of femininity (and masculinity) also have other, unintended side effects. Because they still maintain a kind of femininity (and masculinity) as represented by Barbie (and Ken) in the public sphere and allow these to radiate and seduce (Schober 2009:316), this can often produce very conflicting emotional responses. One

Picturing "Gender" 73

example occurred in connection with one of the above-mentioned election posters placed near the University of Vienna. Someone covered it with angry graffiti, "I love Barbie. Just because you wear flat shoes doesn't mean you're independent. Shitty leftist feminist ball breakers! ©A woman with tits." In this statement Barbie is not addressed as a norm to be combated but as an

Figure 5.4   Graffiti, Berlin, Friedrichshain, 2009, © Henning Onken.

ideal with erotically seductive features and libidinal connotations: she is connected with "love" as well as "tits" and this way counter-posed to feminist, emancipatory discourse which is associated with un-erotic "flat shoes."

Whereas in this example feminist discourse is attacked and used as a trigger for articulating more conventional fantasies of erotic femininity, the opposite can also happen: gender then becomes involved in figurations in order to expose (usually not publicly articulated) anger and fury in respect to the reigning body regimes. This is the case in the repeated "kill your gender" or "gender sucks" graffiti that can be found in the streets of Berlin.[22]

Occasionally such a questioning is also attained by the enunciation of a taboo such as representations of bulimia in connection with the quotation of gender stereotypes. Other examples are the representation of a self-perception of strikingly "beautiful," that is, carefully and colorfully staged obese bodies often integrated in very individual and trashy lifestyle settings. The angry writing on the Barbie poster, the repeated "kill your gender" or "gender sucks" graffiti, and the representation of bulimia or of the beauty of obese bodies all demonstrate that popular figurations of gender are temporarily able to accommodate often heated feelings, sensations, and experiences that lie beyond what is explicitly represented in official discourse in respect to our life as bodily beings. These figurations then function as "settings in which knowledge that is normally unspeakable can be articulated" (van de Port: 1998:101).

New Pugnacious Everybodies: Female Migrants and the Clown

It is striking that where political mobilization is an issue, it is not images that display symmetry or images that show a self in metamorphosis or attacks onto body norms that are employed, but the asymmetry of power is again highlighted and universality is again embodied via new particular-universal figures. One such example of new "everybodies" (de Certeau 2002:2; Schober 2001:233) addressing us as an audience are figures of the (female) migrant which fractions of the alter-globalization movement often use in order to popularize their messages.[23] Like older figures of this type, such as "the worker" or "the woman," this female migrant addresses "all of us" and stands for truth claims in respect to what is discursively represented. This figure makes it possible to expose a gap between current asymmetries of power and a potential equality and parity. But like older everybodies such as the worker, they also allow a mirroring and constituting of the protesting and acting self in the image of the other and in doing so can romanticize or "swallow" the other for a repositioning of the self.

Sometimes, however, in relation to gender a new figure assumes this function of an "everybody" in the political arena and somehow bridges the convention of the sexually indistinctive, androgynous self and that of "older" emancipation and social-reform discourses and their everybodies. Then the figure of the (female) clown is employed in order to fulfill this function.

The clown is another central figure related to the concept of gender. In political activism there is the "clown army" of the alter-globalization movement, which sees the clown as a kind of "third gender" and as a symbol of the pacifist able to churn up fixed friend-enemy categories. The hip-hop clowns and "clowning" (versus "krumping") are performed in order to throw black/white dichotomies out of balance. And in art there are the "Paradise Portraits" by the Dutch artist Erwin Olaf (2001) and the clown series by Cindy Sherman (2004).

The mask of the clown is closely related to the audience. It operates more via the body and gestures than via speech and is intended to provoke laughter and to challenge and involve the public, but at the same time she or he is often frightening. The clown usually appears as an insular figure—even within a group of clowns each one appears as an individual. Furthermore, the clown's mask is fixed, somehow "frozen." Difference and melancholic individuality unfold within, something with which we as the audience are invited to empathize via a concentrated examining of gestures and mimicry (Barloewen 2010). Besides, there is something grotesque in the mask of the clown, which also has to do with an undefined, strangely digressing sexuality, which tends toward the asexual, but is again oriented on masculinity. Because the clown is generally a male, autonomous, and lonesome figure, the female clown is the alteration of a male mask.

Similar to the female migrant, the clown is also an ambiguous everybody. The mask of the clown obliterates differences, affirms fantasies of uniting the male and the female into one "unisex" person (oriented on masculinity), and accentuates the interior life of judgments, feelings, and ambivalence. But simultaneously a relating with the clown allows people today to unite with other political activists and to create a public presence for certain claims. This mask, however, also highlights an awkwardness associated with this kind of activity today. The clown no longer has the security and certainty that older everybodies had (such as the worker), but tends instead to expose reflexive self-questioning and a clumsy stumbling.

CONCLUSION

It was not my intention in this chapter to break the various image creations in connection with gender or sex-gender down into a single image or to extract one strand of narration from them, but rather to track the course of these concepts through the public sphere and to highlight not only the ambiguities but also the struggle and conflict that arise in connection with them. Nevertheless, the following types of image acts seem to predominate in connection with these concepts:

- Those that assert, promote, and authenticate a symmetrical co-presence of the male and the female

76 *Anna Schober*

- Those that highlight the singular body and its potential to collect sexual and erotic codes of both sexes
- Those that celebrate a creative bricolage of the patchwork self
- Those that expose a (deconstructive) engagement with incorporations of norms but at the same time maintain figurations of such norms

A further uniting feature of several of these image acts is that any form of asymmetry, for instance, an asymmetry in access to public power but also asymmetries in imaginary projections vis-à-vis the "other" sex, seems to recede into the background. Moreover, male and female body parts and their re-combinations are mainly used as positive tropes, are emphatically staged, and linked to fantasies of control. Any painful, problematic, disordering, confusing, and unsuccessful attempts of body and self-modification are usually not represented and thus tend to disappear. But at the same time artists or graffiti activists sometimes express anger, fury, and frustration by linking up with gender or allow ambivalent feelings and things that are usually not speakable to develop and be articulated in respect to current body regimes. And new sites of staging claims and making asymmetries an issue, such as the figure of the female migrant or the clown, are also created.

In conclusion, images and bodies that arise in connection with gender affirm a kind of vision of the public sphere as being (already successfully) egalitarian and populated by discrete bodies that interiorize ambivalence and difference. At the same time gender is also translated into figurations (such as the female migrant or the clown) that question and challenge this vision. With the latter gender is used to disrupt the security of representing a neutral and universal collective subject, but at the same time the picturing of gender tends to neutralize inequalities in themselves because it usually conceals asymmetries and conflicts between the sexes.

NOTES

1. In this text I do not intend to arrive at one single, fixed definition of the meaning of "gender," nor do I aim at developing a new theory of gender. On the contrary, I am concerned with the ways in which gender appears in the public sphere and the transnational adoption processes and struggles that this involves. Gender thereby serves as a conceptual reference point that can include various cultural constructivist definitions, which are not always explicitly specified but are sometimes referred to via allusions or in an implicit way.
2. The project presented in this chapter received funding from the EU, Seventh Framework Programme, Marie Curie Actions, Grant Agreement No. PIEF-GA-2009–234990.
3. In relation to discussions about "European integration" Andreas Langenohl (2006) has criticized the scientific practice of opposing "culture" and the "political public sphere."
4. A first "cultural" definition of the "sex-gender system" was provided by Gayle Rubin (1975). Judith Butler adopted and expanded several of the concepts

developed by her, but criticized her non-cultural (and guided by utopia) definition of sex as non-cultural (Butler 1990:41).
5. "Third-wave feminism" is used for several diverse strains of feminist activity and study that began in the early 1990s and emerged as a response to perceived failures and a backlash against initiatives created by earlier feminist activity (e.g., during the 1970s).
6. On this notion of "hegemony," see Laclau and Mouffe (1995:96).
7. On the history of the various shifts in this discourse, on an international level but in particular related to discussions in Germany, see Hark (2005). On the differences between various local milieus throughout Europe and Russia in this respect, see Braidotti (2002).
8. For example, Jürgen Habermas' view of the public sphere is guided by an assumption of rationality. For a critique of Habermas, see Calhoun (1992).
9. This outstanding position that an occupation with norms has for most strands of gender studies is described (and affirmatively adopted) by Klaus Rieser (2006:41).
10. On traditions of this binary tendency of judging, see Schober (2010).
11. This does not mean that I think that gender cannot be redefined in a way that makes it useful for contemporary knowledge production—but this is not the aim I pursue in this chapter. I am only concerned with the public appeal and the public life of images invented around gender as they have been in practice for approximately two decades. Such an analysis of the public and political appearance and praxis of gender can provide a basis to rethink and to re-evaluate the use of gender for future emancipation strategies.
12. Since 2009 this image has been featured on several websites and blogs in relation to various types of gender issues. For instance, http://www.myteenagewerewolf.com/lauren/vive-la-difference/attachment/gender/ (retrieved July 10, 2011).
13. A more extensive analysis of such a representation of symmetry in connection with gender can be found in Schober (2011).
14. See, for example, http://merricherri.blogspot.com/2008_07_01_archive.html (retrieved July 10, 2011).
15. Although in interviews in connection with this painting she refers to a transvestite with "a natural penis and false silicon breasts" she was working with (interview with Schama 2005:126).
16. Chris Straayer has shown that performances in music videos since the 1980s show a similar hypersexualization by collecting codes of femininity and masculinity in one body. She calls this "She-Male-Performance" (Straayer 1996:88).
17. This mythic depiction of the androgynous body has a kind of "healing" function since it invites the beholders to a position far beyond the lack that characterizes every identity. Furthermore, in this example the androgynous is not depicted in a merely spiritual way (as the unification of two opposing principles), but in an emphatic bodily way. We have a penis (linked to the phallus—man's societal power) and not a vagina, that is, the proudness and "wholeness" of the presented body is juxtaposed with what in psychoanalytical terms has been discussed as the notion of "lack" with regard to the vagina and women. At the same time the myth of the phallic woman (which is also the castrating woman) is also evoked. This allows for a variety of similarities as well as differences in reading processes of male, female, gay, or transsexual beholders.
18. A related image cluster in relation to gender can be summarized as "the asexual" and represents the "other face" of the hypersexual in which sexual organs or erotic codes are largely missing or appear in an "abject" way. An

example for this are the installations of childlike and physically "sexless" creatures by Jake and Dinos Chapman (Mahon 2005:282).
19. Saville "gets bag loads of post every month from fat women who are pleased she has recognised their beauty" (Milner 1997:4).
20. For instance, http://www.queeryouth.org.uk/community/index.php?act=ST&f=8&t=23044 (retrieved June 1, 2010) and http://jessicaivins.net/bodywatch/artwork.html (retrieved June 1, 2010).
21. Linda Nochlin (1994:8) analyzes the history of the body fragment as a positive and even revolutionary trope in the context of modernity.
22. http://www.fensterzumhof.eu/bilder/v/Berliner-Streetart/berlin-graffiti-gender-kill-sexismus.html (retrieved July 10, 2011).
23. http://www.adbusters.it/pages/serpica.php (retrieved July 10, 2011). On the use of these images, see also Mattoni and Doerr (2007).

REFERENCES

Barthes, Roland. 2008. *The neutral: lecture course at the Collège de France (1977–1978)*. New York: Columbia University Press.
Bauman, Zygmunt. 1991. *Modernity and ambivalence*. Cambridge: Polity Press.
Braidotti, Rosi. 2002. "The uses and abuses of the sex/gender distinction in European feminist practice." Pp. 285–307 in *Thinking differently: a reader in European women's studies*, edited by G. Griffin, and R. Braidotti. London: Zed Books.
Butler, Judith. 1990. *Gender trouble*. London: Routledge.
Butler, Judith. 2004. *Undoing gender*. London: Routledge.
Calhoun, Craig, ed. 1992. *Habermas and the public sphere*. Cambridge, MA: MIT Press.
Castoriadis, Cornelius. 1997. "The discovery of the imagination." Pp. 213–245 in *World in fragments: writings on politics, society, psychoanalysis, and imagination*, edited by D. A. Curtis. Stanford, CA: Stanford University Press.
Certeau, Michel de. 2002. *The politics of everyday life*. Berkeley: University of California Press.
Damisch, Hubert. 1979. "Eight theses for (or against?) a semiology of painting." *Enclitic* 3(1):1–15.
Deitch, Jeffrey, ed. 1993. *Posthuman. Neue Formen der Figuration in der zeitgenössischen Kunst*. Hamburg: Deichtorhallen-Ausstellungs-GmbH.
Didi-Huberman, Georges. 1992. *Ce que nous voyons, ce qui nous regarde*. Paris: Les Éditions de Minuit.
Fraisse, Geneviève. 1996. *Geschlechterdifferenz*. Tübingen: Edition Diskord.
Francblin, Catherine. 1992. "Cindy Sherman, personnage très ordinaire." *Art Press* 65(January):12–19.
Hark, Sabine. 2005. *Dissidente Partizipation. Eine Diskursgeschichte des Feminismus*. Frankfurt/ Main: Suhrkamp.
Hegener, Wolfgang. 2009. "Die Ambivalenz des Ursprungs. Diesseits und jenseits von Geschlechterdifferenz und Sexualität." Pp. 129–147 in *Postsexualität. Zur Transformation des Begehrens*, edited by I. Berkel. Gießen: Psychosozial Verlag.
Helfand, Glen. 1997. "Cindy Sherman: from dream girl to nightmare alley." Retrieved August 30, 2012 (http://www.salon.com/1997/12/08/media_38/).
Jaggar, Allison. 1983. *Feminist politics and human nature*. Totowa, NJ: Rowman & Allanheld.
Kallfelz, Andreas. 1984. "Cindy Sherman: 'Ich mache keine Selbstporträts'" *Wolkenkratzer Art Journal* 4 (September–October):44–49.

Kinroos, Robin. 1989. "The rhetoric of neutrality." Pp. 131–143 in *Design discourse*, edited by V. Margolin. Chicago: University of Chicago Press.
Laclau, Ernesto and Chantal Mouffe. 1985. *Hegemony and socialist strategy: towards a radical democratic politics.* London: Verso.
Langenohl, Andreas. 2006. "Öffentlichkeit und politisch-kulturelle Differenz in Europa: Jenseits von Kulturalismus und Anti-Kulturalismus." Pp. 177–196 in *Demokratisches Regieren und politische Kultur. Post-staatlich, post-parlamentarisch, post-patriarchal?* edited by K. Ruhl, J. Träger, and C. Wiesner. Münster: LIT.
Lupton, Ellen. 1989. "Reading Isotype." Pp. 145–156 in *Design discourse*, edited by V. Margolin. Chicago: University of Chicago Press.
Mackenzie, Suzie. 2005. "Under the skin." *The Guardian*, October 22.
Mahon, Alyce. 2005. *Eroticism and art.* Oxford: Oxford University Press.
Marin, Louis. 2007. *Von den Mächten des Bildes. Glossen.* Zurich: diaphanes.
Mattoni, Alice and Nicole Doerr. 2007. "Images within the precarity movement in Italy." *Feminist Review* 87:130–135.
Meagher, Michelle. 2003. "Jenny Saville and a feminist aesthetics of disgust." *Hypatia* 18(4):23–41.
Mersch, Dieter. 2007. "Blick und Entzug. Zur 'Logik' ikonischer Strukturen." Pp.55–69 in *Figur und Figuration. Studien zu Wahrnehmung und Wissen*, edited by G. Boehm, G. Brandstetter, and A. von Müller. Munich: Wilhelm Fink.
Milner, Catherine. 1997. "Bring on the blubbernauts." *The Sunday Telegraph*, September 14:9.
Mitchell, J. W. T. 2005. *What do pictures want? The lives and loves of images.* Chicago: University of Chicago Press.
Mulvey, Laura. 1996. "Cosmetics and abjection: Cindy Sherman 1977–1987." Pp. 65–76 in *Fetishism and curiosity*, edited by L. Mulvey. Bloomington; London: Indiana University Press; British Film Institute.
Nochlin, Linda. 1994. *The body in pieces: the fragment as a metaphor of modernity.* London: Thames & Hudson.
Putino, Angela. 1998. *Amiche mie isteriche.* Napoli: Cronopio.
Rieser, Klaus. 2006. *Borderlines and passages: liminal masculinities in film.* Essen: Die Blaue Eule.
Rubin, Gayle. 1975. "The traffic in women: notes on the 'political economy' of sex." Pp. 157–201 in *Toward an anthropology of women*, edited by R. Reiter. New York: Monthly Review Press.
Schama, Simon. 2005. *Jenny Saville.* New York: Rizzoli.
Schober, Anna. 2001. *Blue Jeans: Vom Leben in Stoffen und Bildern.* Frankfurt/Main: Campus.
Schober, Anna. 2009. *Ironie, Montage, Verfremdung. Ästhetische Taktiken und die politische Gestalt der Demokratie.* Munich: Wilhelm Fink.
Schober, Anna. 2010. "*Undoing Gender* revisited: Judith Butler's parody and the avant-garde tradition." *Gender Forum. An Internet Journal for Gender Studies* 30. Retrieved October 17, 2012 (http://www.genderforum.org/issues/de-voted/undoing-gender-revisited/).
Schober, Anna. 2011. "Gender und (A-)Symmetrie." *Zeitschrift für Kulturphilosophie* 5(2): 377–400.
Shaw, Jo. 2002. "The European Union and gender mainstreaming: constitutionally embedded or comprehensively marginalised?" *Feminist Legal Studies* 10:213–226.
Straayer, Chris. 1996. *Deviant eyes, deviant bodies: sexual re-orientations in film and video.* New York: Columbia University Press.
Stratigaki, Maria. 2005. "Gender mainstreaming vs. positive action: an ongoing conflict in EU gender equality policy." *European Journal of Women's Studies* 12(2):165–186.

Taylor, Paul.1985. "Cindy Sherman." *Flash Art* 124(October–November):78–79.
van de Port, Mattijs. 1998. *Gypsies, wars and other instances of the wild: civilization and its discontents in a Serbian town*. Amsterdam: Amsterdam University Press.
von Barloewen, Constantin. 2010. *Clowns. Versuch über das Stolpern*. Munich: Diederichs.
Wittgenstein, Ludwig. 1984. *Werkausgabe*. Vol. 8. Frankfurt/ Main: Suhrkamp.
Zerilli, Linda M. G. 2005. *Feminism and the abyss of freedom*. Chicago: University of Chicago Press.
Zimmermann, Tanja. 2006. "Bild-Sprachen fuer neue Menschen: Ideogramme und Pictogramme im revolutionären Russland." Pp. 38–57 in *Piktogramme—Die Einsamkeit der Zeichen*, edited by M. Ackermann. Munich: Deutscher Kunstverlag.

Part II
New Methodologies for Sociological Investigations of the Visual

6 Production of Solidarities in YouTube
A Visual Study of Uyghur Nationalism[1]

Matteo Vergani and Dennis Zuev

INTRODUCTION

The uprising in Tibet, which gained the attention of the global media during the preparation for the Beijing Olympics in 2008, and the ethnic riots in Xinjiang in 2009 showed Western audiences the extent of the ethnic conflict between China's Han ethnic majority and its ethnic minority groups, the Uyghurs[2] being one of them. In the late twentieth century, Uyghur diaspora communities formed their own cultural and political associations in Central Asia, Russia, Turkey, Europe, and North America, while Uyghur political activists have been increasingly using the Internet for spreading political messages promoting Uyghur national ideas. Among their efforts, different Uyghur organizations and individuals have created YouTube channels in order to generate interest in the Uyghur question, to internationalize Xinjiang[3] issues, and to seek support from a variety of audiences.

The aim of this research is to examine the structure of political communication about the Han–Uyghur conflict on YouTube, taking into consideration the contents of the videos and available information about authors and audiences. This study combines quantitative and qualitative methods of data analysis and offers insights into how the Internet is used by dispersed political actors within a framework of a single nationalist movement. It also provides a unique window into ethnic relations in Xinjiang and elucidates on how ethnic minority and transnational activist networks search for discourses that could serve as the basis for ethnic mobilization.

CONTEXTUALIZING THE NEW FORM OF UYGHUR NATIONALISM

Today, Xinjiang is one of the most politically sensitive areas in China, and along with Tibet it is the locale where "preferential policies" toward minorities are in effect. These policies are assessed by various scholars both in positive and negative manners.

For instance, Barry Sautman (2005:105) argues, "If preferential policies did not exist, tensions would likely be higher than they are at present because avenues for maintaining aspects of minority lifestyles and for advancement into the middle class would be closed." Two of the important economic measures in the policy on Xinjiang have been preferences in fiscal transfers from the central government and ability to keep 80% of tax revenue against the usual 50%. Therefore, according to these figures, Xinjiang cannot formally complain about a colonial attitude by the center, as historically it has not been exploited but even greatly subsidized (Christoffersen 1993).

China's Great Western Development plan, however, entailed assimilation and integration of Uyghurs (Clarke 2007)—the price to pay for future economic well-being. Two main causes of societal insecurity in Xinjiang are forced cultural integration and the shift from discrimination to segregation observed by Becquelin (2000). In fact, as Sautman (2005:104) observes, the "erosion of minority cultures" is the issue which remains unaddressed by Chinese ethnic policymakers. The threat to the collective identity of ethnic minorities becomes the core reason for ethnic mobilization and resistance expressed by diverse means: either violent, as was demonstrated by the riots in Urumqi in 2009, or soft discursive ones (Baranovich 2003; Bovingdon 2002), specifically by devising a new language of democracy (Chung 2006), which will be appealing to western audiences and actors seen as potential sponsors and supporters of the Uyghur agenda.

Because in Xinjiang political resistance in the public domain is prohibited, it is confined to cultural realms (Bovingdon 2002), and besides adopting a Westernized language of democracy, a new generation of Uyghurs finds a different political opportunity for resistance—in the cyberspace. YouTube videos become a new medium of soft or non-violent discursive "infrapolitics," a channel of resistance against the formal representations of social and political life. Videos, however, often get decontextualized in hybrid media spaces and become decoupled from options to act (Jenkins 2009). Because the videos produced by the Uyghurs on YouTube serve as political communication and are embedded in the frame of the Uyghur nationalism and history of Han–Uyghur ethnic relations, this decontextualization certainly decreases the potential effect of original messages in the videos.

The case of Uyghur nationalism is interesting for its combination of religious and ethnic aspects. The roots of pan-Turkist nationalism in Xinjiang date back to the beginning of the twentieth century and can be attributed to the trends which emanated from the Ottoman Empire (Waite 2006). With the independence of Central Asian republics, pan-Turkist nationalism in Xinjiang was invigorated by the growth of pan-Turkist sentiments in Central Asia. One of the important nuances in this trend was the fact that Uzbekistan was the most fertile ground for pan-Turkist proclamations (Kubicek 1997) with Uzbeks being the linguistically closest group to Uyghurs. Pan-Turkism as an alternative secularist ideology to Islamic fundamentalism has been considered even more potent as the unification base for the Central Asian states. As we will demonstrate later, this particular ideology has the richest symbolic resources and is most likely to become the central one in the Uyghur nationalist movement's visual and oral discursive repertoire.

A SOCIAL DEFINITION OF YOUTUBE

As the focus of the chapter is YouTube videos, it is necessary to explain what we mean by defining YouTube as a social space. First of all, YouTube is "a Web 2.0" website where the Web 1.0 typical ethos of "build it and they will come" is being replaced by one of "they will come and build it" (Birdsall 2007). It is a space where users participate, socialize, define, and share social norms, sometimes even generating so-called "communities of practice" (Wenger, 1998). Without the active participation of users, Web 2.0 services, such as YouTube, Wikipedia and Facebook, would totally lose their significance. Second, YouTube is not only a video sharing website but also a social network website (Boyd and Ellison 2007; Jenkins 2007). YouTube not only allows the user to create a "channel" (a personal page to upload videos and personal information) and to articulate several lists of linkages (friends, subscriptions, subscribers, favorite videos); by facilitating sharing practices YouTube in fact allows users to maintain and strengthen individual as well as group identities (Lange 2007). Through the sharing of personal videos on YouTube it is possible to reconnect and to enhance family ties among people living far away, creating a sense of intimacy (Rouse 1991). The publishing of videos and comments is an activity that not only allows people to consolidate existing (or smashing—see, e.g., Rogowski's chapter in this volume) social relations, but also to create new ones through the web service, for instance, among political activists.

According to Lange (2008), not only does YouTube allow the germination of communities, but YouTube itself is also often perceived as a community by its users. Many people watch YouTube videos and move on, without spending much time on the site; yet others easily spend hours watching YouTube videos, commenting on them and interacting with other users. YouTube has many mechanisms for drawing people into the "video vortex" of watching videos in a participatory manner.

Finally, the features of YouTube communication recall Castells' concept of "mass-self communication" (Castells 2007). On the one hand, the website is a communicative arena in which users create a variety of social groups elaborating identities through the sharing of symbolic representations. On the other hand, YouTube is a social space frequented not only by members of communities, but also by other people in different times, places, ways, and contexts of fruition.

86 *Matteo Vergani and Dennis Zuev*

This chapter presents a field research based on the use of both qualitative and quantitative methodologies. The quantitative analysis gives some contextual and general insights on the following questions regarding the production of Uyghur videos on YouTube: How many videos have been uploaded? From which country? In what language? What kind of images they use? What genre is the most represented? The data presented and analyzed in the following pages will be further elaborated on in the qualitative part of the chapter where we will also present a typology for the analysis of the visual and ideological content of the videos. Finally, in the last section of this chapter we will integrate a discussion on the data coming from both the qualitative and the quantitative methods, and we will conclude with our final remarks.

QUANTITATIVE STUDY: METHODS AND TECHNIQUES

In order to produce reliable quantitative data about all the YouTube videos on Uyghurs (available at the moment of conducting the research), we chose to retrieve the whole list of videos about Uyghurs using YouTube's search engine and then to extract a statistical sample for our analysis. To overcome the obstacles produced both by the folksonomy (Golder et al. 2006; Vergani 2011) and by some YouTube search engine bugs (such as duplicates in the lists), we developed an original technique consisting of the following:

- Collecting a list of keywords referring to the Uyghurs. Using the Atlas.ti software for conducting qualitative textual analysis, we analyzed several texts written by various stakeholders (such as Uyghur diaspora associations, Turkish, European and American political groups supporting the Uyghur cause, radical Islamic groups supporting Uyghurs, Chinese political leaders attacking the Uyghurs). The result of this analysis was a list of all the different ways of referring to the Uyghur conflict: we found 321 keywords (e.g., Hasan Mahsum, Rabbiya Kadeer, Xinjiang, Uyghurstan, East Turkestan, and so on). We used these keywords for setting the search up on YouTube. It is our assumption that these keywords were the most common words used for defining the Uyghur–Han conflict.
- For each search we used a different keyword. However, we did not restrict the videos to the list showed by the YouTube search engine. From each of the videos, we browsed to the author's YouTube Channel page, looking for other videos about Uyghurs within the channel, both looking at the videos of the same author and at the videos of the author's friends and subscribers. Using this technique, after about twenty searches we reached a "saturation" point in which all the authors of Uyghur-related YouTube clips were already included in the list.

At the end of this exhaustive process, we collected a list of 5,470 videos about Uyghurs on YouTube (December 2009). For our analysis we extracted a statistical sample of 270 videos using a random number generator and analyzed the videos using a coding sheet focusing on the following aspects:

- Features of the content of the videos. During a preliminary viewing of Uyghur videos on YouTube we identified several different genres: cultural (videos showing traditional music, dance, and folk legends); political (videos explicitly exposing political and ideological issues on the Uyghur affair); homemade tourist videos of holidays; fiction movies (pieces of Uyghur movies taken from TV broadcasts); funny and satirical sketches; information programs; personal private videos about friends and daily life; and advertisements. In addition to that, we identified the main language used in the videos (Uyghur, Chinese, Turkish, English, or other). We tried to understand which genre of videos about Uyghurs was the most represented on YouTube, and in which language.
- Personal information about the author: country, age, and gender. In some YouTube channels there is a considerable amount of private information about the author, while in others the authors remain essentially anonymous.
- Technical issues about the production of the video, such as the origin of images (mainstream media, user generated content, or a mix), the style (slideshow presentation, or a motion video) and the duration of the videos.
- Paratextual information available on the channel page (number of videos uploaded by the author; number of friends and subscribers), and on the page of each of the videos (number of comments and views; uploading date).

THE OUTCOMES OF THE QUANTITATIVE RESEARCH

The most represented genre of Uyghur videos on YouTube is definitely cultural (61.5%), and within this category musical videos are the unquestioned majority. Videos about Uyghur traditions, folklore, dances, and weddings are the overwhelming majority of all the other videos in the cultural genre. The second genre of relevance is the political genre, counting 12.2%, that is, about one fifth of the cultural videos. The funny and satirical videos count 8.9% of the sample, while videos containing information programs represent 5.9%. The homemade tourist videos are 4.4% of the sample, fiction movies are 3.3% of the sample, and private videos with images of friends and family are 3% of the sample. Advertisements are only 0.4%, with only one case. These percentages mean that there is a great availability of multimedia contents about the cultural traditions of Uyghur. It follows that by searching YouTube for Uyghur videos one will most likely find a video showing a Uyghur cultural element.

When examining the characteristics of Uyghur-related video uploads we found an increase in uploading videos during the pre-Olympic and Olympic period (May–August 2008). This is probably due to the international attention about China and its political situation. We also found that on average, Uyghurs from European countries upload a smaller percentage of cultural videos (36.8%) than the total average (61.5%). The majority of videos uploaded by Uyghurs based in Europe are political videos (26.3%). We noticed the same trend among the Uyghurs in Turkey. While the total average of political videos is 12.2%, authors identifying themselves as coming from Turkey upload 22.2% of political videos. People identifying themselves as coming from China upload just a few humorous or satirical videos (1.9% relatively to the overall average of 9%), and no political videos at all. It is clear thus that in Europe and Turkey there are Uyghur diaspora communities who are more involved in Uyghur politics than anywhere else.

Overall, more than 50% of the sample comes from authors who identify themselves as coming from the United States and China; about 25% come from Turkey, and only 7.1% from Europe. Only 8.6% of the authors do not indicate their country, while over 40% do not reveal their gender or age. Even if it is possible to lie about personal information, these numbers show how country is considered a more public aspect (or maybe politically more important for the authors) than age and gender.

With regards to the languages used in the videos we examined, 53% of the videos analyzed are in the Uyghur language (a Turkic dialect in Arabic script), 13.7% are in Chinese, 12.6% in English, and 7.4% in Turkish (Figure 6.1).

This massive use of the Uyghur language in videos, compared to that of the authors' country of location (China in only 26.1% of the cases), confirms the idea that the majority of the videos are addressed to Uyghur diaspora communities. They are the only ones who understand the Uyghur language. In 78.3% of the cases, the Uyghur language is used for cultural content (especially music videos), and only in 3.5% for expressing explicit political messages. The languages used for political messages are mostly English (39.4% of political messages) and Turkish (30.3% of political messages). Videos where political messages are expressed in Turkish represent half of the videos where Turkish is used. These data mean that the political messages are mostly addressed to a global Western audience (39.4%) and to a national Turkish audience (30.3%). A total of 51.5% of the political videos present circular contents (i.e., an original mix of mainstream content from media such as images collected on the web, with original subtitles and a soundtrack); 27.3% of the political videos are user-generated contents, and only 21.2% are mainstream videos from cinema or TV. The use of mainstream content from other media for the YouTube videos is also widespread in the cultural videos (73%) where most of the content is taken from the Xinjiang regional

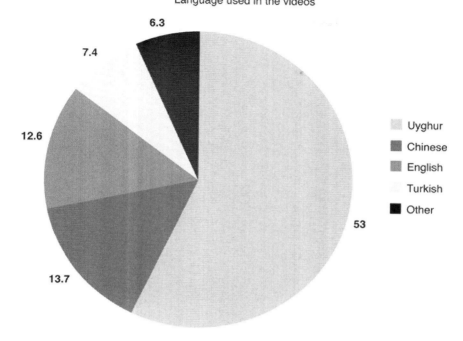

Figure 6.1 Language used in YouTube videos about Uyghurs.

television (XJTV). This means that only a few users present their own original visual content, while the majority of the amateur users (we can define them as non-professional users, so users who do not create videos as a profession) still take many images from the mainstream media, and manipulate this adopted visual content in original frames, creating new contents coming from the mix between mainstream images, amateur slides, and written text.

Only 9.7% of the authors uploaded fewer than five videos onto their channels. The majority of authors (67.1%) uploaded between 5 and 250 videos (a number that can indicate the presence of an active user). There are also many (19.4%) authors uploading a considerable number of videos (between 250 and 1,000). They are very active users, spending a lot of time on the website. Yet political videos are generally uploaded by less active users: 24.2% are uploaded by people having fewer than five videos on the channel, and 45.5% are uploaded by people having fewer than fifty videos on the channel. It is also possible that, in order not to be tracked down by the authorities, people who are uploading political videos create a number of YouTube accounts for that purpose, using different usernames.

Examining the strength of social networks around Uyghur-related video authors, we found that it is not significantly correlated with the genre of the videos they upload. Usually, video authors have a significant number of friends and subscribers to their channel. More than 50% have over than 50 friends and subscribers, and 20% have between 200 and 1,000 subscribers (mostly Uyghurs, according to our understanding of the nicknames they use).

Finally, 46.7% of YouTube videos about Uyghurs have gotten more than 1,000 views. This percentage again shows the existence of a Uyghur community looking for videos and news about their nation, not only on YouTube but also, more generally, on the web. Because YouTube videos are often embedded in Uyghur diaspora websites (as information centers or homepages of activists associations), many of the views are mediated by these websites. The small number of comments that the videos generally receive on YouTube corroborates our assumption. The use of different languages (English, Turkish, Uyghur, and Chinese) in the analyzed sample of the videos shows that different language audiences (see Figure 6.1) participate in production of videos and thus, bring in their models for "preferred reading" (Hall 1980) of the video clips. At the same time, we found videos with comments written in different languages and alphabets, showing the boundary-free effect of the visual contents and the secondary meaning of verbal codes to the visual ones.

THE OUTCOMES OF THE QUALITATIVE RESEARCH

The qualitative analysis of visual materials collected through YouTube or other user-generated content platforms is a nascent field of research (see, e.g., Wachowich and Scobie 2010). Of all available categories mentioned in the previous section, for our qualitative study presented here we chose to focus on the political videos produced by activists of globally dispersed Uyghur diaspora. As mentioned above, our quantitative analysis shows that political videos made up a fair share of the Uyghur YouTube videos and were second only to "cultural" or culture promotion videos. In order to get a better understanding of the visual representations of ideologies and discursive positions held by the main stakeholders in the Uyghur–Han conflict, we developed a typology of the political videos' content. In our analysis we employed the concept of ideological code based on the definition of ideology provided by Hodge and Kress (1993:6): "ideology is the systematic body of ideas organized from a particular point of view." Identification of ideological codes is the analytical strategy which helps us to systematically sort the YouTube videos related to Uyghur nationalism and distinguish prominent sets of meanings, which are incorporated with values, social agents and processes that produced them.

In the next section we outline the means by which ideological codes are related to the videos' interpretive communities.

Reading YouTube Videos: Ideological Codes and Interpretive Communities

In the visual content analysis our task was to identify dominant visual markers (symbols) and ideological codes. Having made a typology of ideological codes, it was possible to analyze the underlying values promoted in the videos and then to pinpoint communities that could be their target audiences. In our study, the concept of "interpretive community"[4] is essential in understanding YouTube video production as a tactical response of Uyghur diaspora to political and cultural hegemony of the Chinese state. It also allows us to approach media audiences in a more organic sense (Lindlof 2002). As Lindlof defines, "an interpretive community is a collectivity of people who share strategies for interpreting, using, and engaging in communication about a media text or technology" (Lindlof, 2002: 64). Interpretive communities are characterized by the use of peculiar visual markers and symbols.

In this chapter ideological codes are associated with two main language communities: Arabic/Turkic and English, both promoting particular values and types of action. We suggest that there are four distinct interpretative communities involved in decoding the messages: (1) a tradition-based community, promoting kinship-oriented values and action; (2) a community based on liberal values, promoting human rights–oriented discourse and action; (3) a community promoting political agenda by means of militant actions; and (4) a community promoting political agenda by means of civil resistance. Each of the communities is visually represented in the videos by certain symbols and mythical representations. For example, kinship is symbolically depicted by the myth of the common origin of the Turks (wolf), and the idea of the free democratic state is depicted by Rabbiya Kadeer, the leader of World Uyghur Congress (a woman politician promoting a free state, freedom of speech, and indigenous rights).

Visual markers make up a conglomerate of symbols, which not only can be used in encoding different ideologies, but may also have unique visual application. For instance, a video of someone with a covered face against the background of a banner with an Arabic script, reading a statement in Uyghur language, and accompanied by a couple of colleagues holding automatic weapons, can be the conventional visual marker of "militant Islam" or "religious extremism." One could argue that ideologies require the use of conventional visual markers or a set of markers for a better persuasion. According to this argument, the visual marker as an exclusive marker can only be used to promote a distinct ideology. However, as we will show in the next section, a particular set of symbols (like the one above) can be used to promote opposing ideologies (in our case: militant Islam and Islamophobia), particularly in cases where the authors of the video are anonymous (thus their organizational and ideological affiliations remain unknown).

In decoding ideological messages, the challenge is to reveal symbolic markers and associations of different ideologies. The four ideological codes that emerged in our analysis—nationalist (pan-Turkist), human rights, militant-political, and civil-political—can be debated in terms of their label accuracy, but they provide an important basis for the analysis of the political videos, particularly when attached to the respective actions promoted in the videos: kinship-oriented action, human rights–oriented action, military action, and civil-political action.

In the process of identifying ideological codes used by YouTube video producers it was not always possible to find a video clearly promoting one particular ideology. For instance, the video called "Dogu Turkistan Gercegi"[5] ("Truth about East Turkestan") employs verbal and visual symbols referring to "genocide"—such as "mutant children" with inflated heads and small arms, nuclear explosions (all these images apparently refer to the Lobnor nuclear test site in Xinjiang). A fair share of the images are made up of Turkish language newspaper headlines such as "Mothers suffering, Horror in China, population planning in China is carried out through killing" or "Chinese are eating East Turkestan children to observe the family-planning policy" (with accompanying images of an Asian man eating something anthropomorphic).[6]

In the next sections we will examine particularities in visual manifestations of the four identified ideological codes.

Nationalist (Pan-Turkist) Ideological Code

In this section we consider the nationalist (pan-Turkist) ideological code as the pivotal one in the structuring of the Uyghur political communication in YouTube. Nationalism in all of the analyzed videos is manifested in two levels: localized aimed at uniting the Uyghurs who live in the People's Republic of China, and in a pan-Turkist level, reaching beyond China's borders, where Uyghurs are seen as one of the many Turkic tribes in the Turan belt,[7] stretching from Bulgaria to Sakha republic in Russia. In the first case, Uyghur singularity or unique Uyghurness is expressed via reference to the photographs of the Golden Ages of Xinjiang and the political success of the Uyghurs in gaining brief periods of independence—during the First East Turkestan Republic (1933–1934) and during the Second East Turkestan Republic (1944–1949). In the other case, common Turkic roots are emphasized by the common past and even by similarity between the Turkish and Uyghur flags. The particular combination of blue and a crescent with a star on the unofficial Uyghur flag is a powerful symbol of nationalism which appears in most of the videos and is present in the videos related to all the four ideological codes. In addition to that, map symbolism is another rhetorical device that serves to distance Xinjiang from China. A frequent image is a red-colored map of non-Uyghur China with Xinjiang highlighted in the color of the Uyghur nationalist blue flag. The color demarcation emphasizes the differences between the ethnic groups of Han and Uyghur (see Figure 6.2). The blue color of the flag recalls the Göktürk (the heavenly Turk Empire) and is also regarded as the flag of Turan, inherited from the colour of Tamerlane's banner (Hyman 1997). The blue functions as the pan-Turkist "civilization" demarcation from the red (which is regarded as the other, a barbarian, a communist).

Two videos, "16 Hunnic Turkic Empires"[8] and "Turan"[9] refer to pan-Turkism and are based on the same scenario. A map of Eurasia is shown in grey and marked as a space dominated by Turkic speaking ethnic groups including Uyghurs, stretching from the Republic of Sakha in the Russian Federation all the way to Bulgaria. Both videos are uploaded by the same user, SultanUyghur, and the maps in the videos contribute to drawing a Turkocentric world with central position of Uyghurs in Central Asia. Figure 6.3 illustrates the extensive use of maps as a distinguishing feature of nationalist (pan-Turkist) code. The maps serve for making introduction into the history of Turkic Empires and their role in state building in Central Asia. In the clip "16 Hunnic Empires" the producer's logo uses the slogan and logo of the Turkish Grey Wolves organization: *Hedef Turan, Rehber Kuran* ("The goal is Turan, the guide is Koran"). The maps of the Turan belt show the Turkic zone of close and distant kinship and help laymen to locate the region on a larger geographic scale.

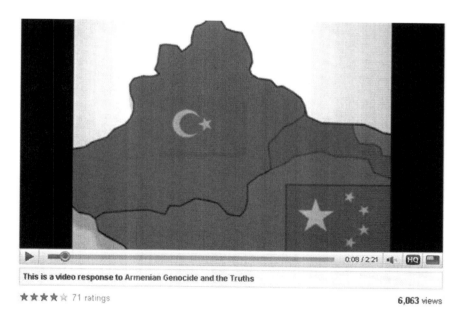

Figure 6.2 Clash of Chinese and Uyghur symbols in nationalist ideological code.[10]

94 *Matteo Vergani and Dennis Zuev*

Figure 6.3 Map symbolism in nationalist (pan-Turkist) ideological code.[11]

Another image in the same clip captioned "16 Magnificent Hunnic, Turkic, Mongol Empires" shows iconic buildings representing three dynasties founded by the Turks at different times: Taj Mahal in Delhi, Gur Emir and Registon complex in Samarkand, and Selimiye Mosque in Edirne. The three buildings represent the architectural heritage of the Turkic Empires and are counter-posed by a red-colored Great Wall of China through which they break (Figure 6.4). The video comprises a sequence of shots highlighting maps of military conquests and the living space expanses occupied by Turkic states and dynasties: Göktürk, Kara-Khanide, Timurid, Mughal, Ghaznevid, and the Ottoman empires. The two main visuals in the video are related with the conflict of the Turkic and Chinese, and this is represented via the juxtaposition of architectural landmarks which stand for both cultures and maps, especially where the Chinese territory is occupied by three consequent dynasties: the Mongols of Genghis Khan, the Golden Horde, and the Yuan dynasty.

Nationalism in the videos we analyzed is manifested not only via conventional symbols such as the flag, territorial boundaries, or iconic buildings, but also by using symbolic personalities which help to create attachment to the ancestors. In the video "Great Uyghur Civilization—Tarim Mummies and Tocharians," which is presented in the format of a slideshow,[12] the producer sends the viewers to the ancient past of the Uyghur culture, which is depicted as equal or even rival to the ancient Han civilization, for instance, by reference to Tarim mummies.[13] Images of Tarim mummies also act to

Production of Solidarities in YouTube 95

Figure 6.4 The iconic clash of Turkic and Chinese civilizations.[14]

support the Uyghur claims of being indigenous to the region, and one of the mummies, named "Beauty of Loulan," is considered a symbol of Uyghur ancient past. After several slides about the mummies, the author invites the viewers to watch the images of "Uyghurs, an Old and beautiful nation." This slide is followed by a row of portraits of children, women, old men, regardless of conventional standards of beauty. The images of common people weave a line between the sanctified objects in the land of ancestors and the vernacular everyday life. The ancient kin and their resting places are related to the present to create attachments and stir the feelings of pride and relatedness to the community.

Some of the videos which are related to nationalist (pan-Turkist) ideological code show the images of "national heroes" of the not-so-distant past. They emerge in a series of black-and-white photographs, while the images of the present "war heroes" or Uyghur culture icons (Hasan Mahsum, head of East Turkestan Islamic Movement; Barat Haji, Uyghur dissident; or untitled portraits of political prisoners executed or to be executed[15]) are followed by dedications in their memory.[16] The video about the folk musician Merhum Koresh Kosen (video "Merhum Koresh Kosen olmeydu"[17]) has no direct mission of organizing any political action or rally. Instead, its main objective is to announce a call to collect all materials about the musician, photographs and video recordings and thereby cement cultural links. The dominating image in this video is the figure of the musician itself, playing *dutar* (Uyghur folk instrument). Thus, the visual imagery here is meant to "awaken" the cultural consciousness and present ways to support the cause or transfer money to a bank account (as in video "Dogu Turkistan Hep Ata Diyari Kalicak").[18]

Rudelson (1997) argues that Uyghur nationalism in Xinjiang is hindered by the multiple local identities and loyalties. However, in our study, we find that the creation of visual links in the YouTube videos between political events of the past, national heroes, and myths, through the cognitive and emotive potential of symbols, makes the pan-Turkist ideology one of the most prominent ideological codes. Considering that the central figure in the Uyghur nationalist discourse, Rabbiya Kadeer, is not a spiritual or religious leader, but in fact a former politician and entrepreneur, makes pan-Turkist ideology the most convenient core of Uyghur nationalism rather than radical Islam.

As we discovered in the videos, pan-Turkist ideology is signified by one or more of the following components: (1) visual representations of symbolic events of the past (such as the establishment of independent Turkestan Republics); (2) symbolic figures; (3) archetypes (such as a wolf, a horse, or a camel) and myths (such as the Aryanism); (4) cultural heritage (music, language, dress); and (5) Uyghur landscape icons (Tian Shan mountains, Sairam lake, Taklamakan desert, vineyards of Turpan, mosques in Kashgar, ancient ruined cities in Tarim basin, and donkey carts). The Uyghur territory becomes less of an imaginative construction than a real, attainable entity, and the symbols of home become an important visual aspect of the ethnic solidarity.

Human Rights Ideological Code

To the Chinese government's sensibilities, the human rights and militant-political ideological codes represent two threat discourses. The videos that fall within the human rights ideological code use "genocide" as a keyword and are explicitly anti-Chinese. They depict the Chinese state as a threat to the survival of the Uyghurs as a nation. The militant-political code is closely connected to the "religious fundamentalism" idea, and thus perceived as a threat to the Han Chinese and the security of the Chinese state. Such videos of "warnings," issued by the different Uyghur separatist ("freedom fighters") to the Chinese government, are perceived as visual goods for the Chinese authorities to legitimize their "war on terrorism" (see Shichor 2006).

The issue of genocide and human rights ideological code in political communication is represented in videos produced by Turkish and Uyghur users. In the videos, signifiers of genocide are associated with the physical damage to the Uyghurs inflicted by Han Chinese. For instance, several videos demonstrate the victims of nuclear tests in Xinjiang (mutant animals or children with birth defects) and present a "thriller" with an English-speaking

Figure 6.5 Images of purity and nature in human rights ideological code.[19]

human rights activist group discovering documentation of a higher cancer rate among the Xinjiang population compared to the rest of China.[20]

The human rights code is additionally manifested by the images of environmental or ecological threat to the survival of the minority. Politicization of environmental issues is not new in ethnic mobilization and has often led to the emergence of successful nationalist movements (Geukjian 2007). In some of the videos we can observe the strategy of visualizing the ecological threat to the survival of the Uyghur landscape by the Han Chinese economic plans in Xiniang. Another feature of the code is depiction of Xinjiang as the region where a clean environment, represented by the green forest and blue river, is threatened by the Chinese industrialization plans (Figure 6.5).[21]

In many cases we found that visual manifestations of ideologies are often combined into imagery complexes. In the human rights code it is an imagery set of pictures of mutant children, nuclear explosions, and verbal statements, such as "Uyghur women are transported to Inner China as slaves and sex laborers" (in a video "Is East Turkestan a Chinese Territory?").[22]

Militant-Political Ideological Code

Along with the politicization of human rights issues, religion is also used in Uyghur ethnic mobilization. Militant-political videos tend to show the intentions of some nationalists to resort to violent measures in resolving the political conflict and gaining independence. Despite the assertions of scholars, that Islam as a basis for building Uyghur solidarity has never been very strong due to the status of Uyghurs as borderland people (Roberts 1998), we have discovered that the Uyghur nationalist movement is also represented by videos with imagery complexes that are associated with the "new religious politics" or militant Islamism. In the videos "Hasan Mahsum" and "Islamic Party of Turkistan: Dunya Musulmanlirigha Omumi Mu"[23] the imagery complexes employ popular inventory of symbols associated with militant Islam: men with long beards, automatic rifles, military training, desert landscape, and written and recited Koranic verses. The videos, however, do not go beyond the popular genre of anonymous warnings, and they do not pronounce the goal of creating some version of Islamic state or emirate (see Figure 6.6). Moreover, the veracity of such videos and the intentions of the producers are questioned by some viewers, as comments to such videos reveal, and may be accepted by others.

Militant-political ideological code is represented by groups of militant activists, usually positioned in some anonymous desert landscape and

Figure 6.6 Militants from Islamic Party of East Turkestan issue a warning to the Chinese authorities.[24]

addressing the audience as they issue warning of military struggle.[25] Militant activists choose not to reveal their identity and to mask their faces so well that it is difficult to guess, whether it is a Uyghur or a Han Chinese masquerading (see Figure 6.6). Some of the videos related to militant-political codes show militants' demonstrative field exercises (but not a real-life shooting of military operations).

The image of a militant Uyghur in the videos related to militant-political code is countered by the civil-political code, which is characterized by the images of civil Uyghur protesters and activists vocalizing at press conferences, meetings, and street demonstrations. In fact, the crucial role in internationalizing the Uyghur nationalist movement is played by civil-political organizations, such as World Uyghur Congress (Chen 2010). In that respect, existence of the militants may considerably hinder the efforts of the civil activists, who are seeking recognition and aid from the Western community, which is scared by the images of the Taliban and Afghanistan-trained groups of independence fighters of any origin.

Civil-Political Ideological Code

If militant-political code is loaded with images of military action and armed struggle, the civil-political code represents the civil activism of diaspora by means of demonstrations, public protests abroad, and other forms of peaceful resistance (including music). For example, in the video "Rabbiya Kadeer in Japan"[26] a Japanese television news segment shows Rabbiya Kadeer, the leader of the World Uyghur Congress, giving press conference to Japanese journalists. Other characteristic features of videos related to civil-political ideological code are images of political liaisons with other political figures who express anti-Chinese sentiments (e.g., photos of Rabbiya Kadeer and the Dalai Lama), and acts of symbolic violence, such as vandalizing Chinese official symbols (burning the flag or cutting out stars from it).

As we have mentioned in the section on quantitative analysis, one of the most popular types of uploaded videos are musical ones. The problem with the analysis of these videos is that they have two equally significant aspects: the oral and the visual ones. Some of the musical clips (such as folk song performances) have features of the civil-political code, although they also have elements of the nationalist (pan-Turkist) ideological code. First and foremost, Uyghur songs are one of the available forms of resistance and literally "raising a public voice" (Baranovich 2003) inside China, while visual repertoire of the clips, constructed through the popular folk elements—dresses, hair-style, decorations, music instruments, and landscapes—create a ground for the general Turkic-speaking audience to relate through recognizing familiar cultural elements.

The ambiguity of the visual messages and possible differences in meaning-making process between two potential interpretative communities—the Chinese language audience and the Uyghur language speaking audiences—is best evidenced in the music clip "Xinjiang yakeshi" ("Xinjiang Is Good"),[27] produced for XJTV, a major television station in Xinjiang, but uploaded by a user registered in the U.S. with the user's channel devoted entirely to Uyghur music videos. In the video, dancers dressed in Uyghur national costumes sing the song "Xinjiang Is Good" (*yakeshi*). The video takes us from an airport to Uyghur girls dancing in green vineyards, and from vineyards to the modern high-speed highway which the male singer praises in Chinese: "Speed highway is so beautiful . . . and Xinjiang transport is wonderful." The female chorus then adds, "Xinjiang petrol is so beautiful. And Xinjiang watermelons are so fresh. We welcome all the friends from different places to come and visit Xinjiang. To develop Xinjiang is a great thing." The video may be a good example of China official discourse on well-being of ethnic minorities in China and ongoing regional development and modernization, or it can be a promotion video for the developing region to attract investors and visitors. Upon a closer look, the imagery of Xinjiang beauty or greatness is comprised of a mix of traditional Uyghur places and modern Uyghur places (the result of state investment plans). The picturesque of the old and modern are linked together and are given the final shout (diagnosis) by the singers: *yakeshi* ("good!"). The video may either look like a panegyric to the Han Chinese modernity, coming into the once *luohou* (economically retarded) northwestern backwaters of China, or may be a visual proof to neo-colonialist policy (if we look from the Uyghur side at the Great Western Development Plan).[28] Why did the Uyghur user download this video? The user himself does not provide any extremely critical comments, but we can find reactions to the video from other viewers. Some of the comments to the video resist accepting the benevolent intentions of the production side (Chinese state) by posing the following questions:

> If China is so eager to say Uigur is part of the nation and they are working together, why can't they make Uighur be an official language of China as well? Why are they singing in Chinese? Why is this video in Chinese tune? (comment by a viewer, Xinjiang Yakeshi)[29]

The viewers ridicule the Chinese attempts to show the cultural integration of Uyghur elements into China, and symbolically this politics of integration is signified by one single Uyghur word being incorporated into the Chinese language song (the word *yakshi*). The icons of Chinese development are not always perceived by the ethnic minorities in China as a sign of progress. In fact, they may be seen as a tool for colonial expansion, extraction of mineral resources, and militarization.[30] The video serves for elicitation of the imbalance in intercultural communication between the Han Chinese and Uyghurs. It is not only the oral register of the song which provokes negative reaction, but also the Chinese visualization of Xinjiang through the images of Chinese technological development that provokes Uyghurs to see Chinese as inferiors, and as interested in economical exploitation of the region rather than in a peaceful coexistence.

To summarize, the ideological frameworks, which are suggested by the video producers, are not always clear in terms of what social action they seek to promote. The intentions of the video producers can easily be misinterpreted within the same interpretive community, and the ideological codes in the video production reflect rather different possible ways of development of Uyghur nationalism. The variety of ideological codes supports the observations of Becquelin (2000) on dispersion of Uyghur political communities and the observation of Culpepper (2012) on ongoing nationalist competition on the Internet. In our study, YouTube is considered as the site for crystallization of the Uyghur nationalist idea, which in the meantime consists of different ways of attaining solidarity. The difference lies primarily in the richness of the symbolic repertoire which the audiences in the Uyghur diaspora may find most persuasive and appealing.

In the following section we suggest that analysis of political communication in YouTube should go beyond analysis of the codes and signifiers and must include also an examination of the technical variables essential in video production.

Analysis of Visual Production

If ideological and verbal contents are one side of the video analysis, another important aspect is the technical side of visual production. In this section we outline questions and answers related to the visual production of the videos, which had appeared during our research. We have found four aspects essential in the analysis of visual production by Uyghur nationalists: technical execution, duration, color resolution, and place of production. These four aspects can serve as a useful toolkit for other researchers working with visual data on YouTube.

Is the video executed as a piece of motion footage or a slideshow?

The niche of political spectacle in Uyghur political videos is filled with videos of militants, where masked warriors with machine guns and the trappings associated with militant Islam create suspense. The motion video also creates participatory effect and increases veracity of the video, while slideshow has a lower status of veracity. However, both motion video and slideshow have equal chances to be spectacular.

How long is the video?

Short duration helps to bring more rhetoric muscle and compress the message, while long pieces of the same scene can be boring and distracting.

Long conversations in Uyghur without action (as in the video depicting the preaching of Hasan Mahsum)[31] can be exhausting and not too informative for non-native speakers and even other Turkic-language speakers.

What is the immediate context of the video production?

The location of the video production is not always that of China ("occupied homeland"). The characteristic feature of civil-political code is the foreign location of the footage. Most of the street action is shot in Germany, the United States, or elsewhere. The location which can be identified as the territory of China is linked to repression (e.g., in a video "Communist China's execution of Uyghur Turks"[32] which shows court hearings over Uyghur radicals). By identifying the geographic background in the videos using civil-political ideological code we see that Uyghur political activists are essentially conducting their work "in exile" and not in China.

What is the color, resolution scheme of the video? Are there other visuals embedded in the video?

The quality of the visual side of the videos is perhaps not as important for the producers as the ideological message. The optical codes and intertextual references are some of the most recognizable elements of the visual structure. Contents that are considered "sensational" with poor resolution may signify that the video is not accessible anywhere else and is a primary source—for example, made with mobile phones by witnesses, thus giving higher credibility and authenticity to the scene and action in contrast to the high-resolution-edited videos of the mainstream media. In the case of the motion videos depicting Uyghur militants (such as "Hasan Mahsum," mentioned earlier), the poor resolution and unedited messages gives them more authenticity, thus they contrast the edited, professional glamour of the studios of the mainstream media. At the same time when we compared the videos depicting Uyghur militants with another separatist movement (in Chechen Republic of Russian Federation) we discovered that the videos from Chechnya had much higher resolution and better quality sound editing (Vergani and Zuev 2012). In fact, most of the motion videos we have found on Uyghur militants and which were included in this analysis had bad quality sound editing and low image resolution. Slideshows, on the other hand, had better quality picture, were made in color, and incorporated black-and-white historical photographs.

To conclude, we wish to underline that analysis of communicative intentions of the YouTube video producers cannot be limited to the level of verbal and visual narratives but has to be also examined in conjunction with the technical side of the visual-oral production, which gives a better understanding of the limitations and advantages of the medium and the visual skills of the producers.

CONCLUSIONS

Our quantitative research shows a very high percentage of cultural videos, focusing on folklore, music, and tradition. On the one hand, this information may corroborate the idea of YouTube as a recreational tool, primarily used for amusement, diversion, and fun. Yet on the other hand, this disproportionate sharing of cultural videos may reveal a more complex phenomenon. The global image of Xinjiang Uyghurs is marked not only by the Chinese state cultural and language policies, which the Uyghurs accuse of being repressive and discriminating, but also by the recent Chinese official discourse characterizing Uyghur nationalist movements as terrorists linked to Al-Qaeda. As Dwyer (2005) reports, after 9/11 there was a shift in Chinese media discourse from "Uyghur separatists" to "Muslim terrorists." The maintenance of the cultural tradition through the sharing of music videos, dance, traditional representations, and so on, is a way for Uyghurs to face this attack on their identity, linking all the communities of the diaspora through traditions and cultural identity, and broadcasting a positive image of the Uyghurs to a wide audience. Together with the more traditional forms of political communication as the political nonmilitant videos, these videos express a form of resistance to the attack on their identities.

The people publishing YouTube videos on the Uyghur cause come from different countries, and participate in different ways in the production of videos. First, Uyghurs based in Europe and Turkey upload more political videos than cultural ones, while Uyghurs based in the U.S. and China upload more cultural than political. Turks are the ones uploading the highest percentage of political videos. These data may help us to identify the location of today's most politically active Uyghur diaspora communities in the world. Second, it is interesting to note that there are no political videos at all uploaded from China. People from China upload almost all cultural videos about Uyghur traditions, dances, and music. This is a clear clue to the shortage of Uyghur's public political engagement within the borders of China: Internet in China is heavily controlled by authorities, and people uploading videos supporting the Uyghur cause may risk being persecuted. Third, it is also interesting to note that while Central Asian cities like Alma-Ata (Kazakhstan), Bishkek (Kyrgyzstan), and Tashkent (Uzbekistan) host the biggest Uyghur communities in the world, only a few YouTube authors come from there. This may be explained both by a lack of political freedom in the Central Asian countries and by the digital divide: the difficulty in accessing the Internet that people have in these countries. The fact that the majority (67.1%) of authors uploading videos about Uyghurs uploaded between 5 and 250 videos on their channels, together with the information that more than 50% have more than fifty friends and subscribers, testify that there is a big, linked community of Uyghurs on YouTube, connected by friendship and subscription linkages. They mainly come from the countries of the Uyghur diaspora in which YouTube is heavily used: Europe, Canada, United States, and Turkey.

Although there is no particular social networking site for Uyghurs, there is an online community of Uyghurs on YouTube, which despite being ideologically heterogeneous resists the hegemonic narratives of the Chinese state. The religious fundamentalism (represented by militant-political ideological code) as a solidarity basis for the Uyghurs may be the wrong direction for the Uyghur nationalist movement, in its search for Western support while the West itself is not highly sympathetic with political Islam. This may explain why some Uyghur nationalist organizations, such as World Uyghur Congress, prefer to frame their actions as a human rights–oriented movement rather than a movement relying on support from global Jihad and military action. Additionally, ethnicity based on kinship or relational proximity as its underlying value, rather than ethnicity based on religious affiliation, becomes the chief appeal for the Uyghur YouTube community.

Despite the significant portion of videos in English and Turkish, the main addressees of the political messages are the Uyghur diasporic communities. As for political interest, it follows that Turkish-based Uyghurs and the Turkish-speaking community may be the prime receivers of the overseas Uyghur political plea for independence. For the dispersed diaspora the Internet is a highly politicized medium as it is the only available communication channel for the creation of spatial togetherness and expression of transnational loyalty.

ACKNOWLEDGMENT

We would like to thank Regev Nathansohn for his generous valuable comments on this chapter.

NOTES

1. The quantitative data analysis for this study was first presented at the XVII ISA World Congress in Goethenburg, 2010. The data also appeared in the *Journal of Contemporary China* (Vergani and Zuev, 2011).
2. Uyghurs are a Muslim Turkic speaking ethnic group living mainly in northwestern China. The conflict with the Chinese government resulted in the creation of a large Uyghur diaspora, living mainly in Kazakhstan, Uzbekistan, Turkey, and Germany.
3. Xinjiang is the name of the administrative division of the territory, given by the Chinese government. Uyghurs use several different names to indicate the same territory, including Uyghurstan, East Turkestan, and Sharqiy Turkestan.
4. The term was originally introduced by Stanley Fish (1976).
5. http://www.YouTube.com/watch?v=4w2sPxJQ87E (accessed May 10, 2012).

6. The image also circulated on the Internet as a part of the anti-Chinese discourse on the website of the Movement Against Illegal Immigration in Russia (www.dpni.org).
7. Turan in the Iranian mythology is a toponym related to the territory inhabited by nomadic tribes north of Iran. The idea of Turan belt as a political union of Turkic speaking states is promoted by the Turkish nationalist organization Grey Wolves.
8. http://www.YouTube.com/watch?v=NcbH2sZ_9JA (accessed May 19, 2012).
9. http://www.YouTube.com/watch?v=22e3frREGlM (accessed May 10, 2012).
10. This image is a screenshot from the video "China's Occupation of East Turkestan. Genocide of Uyghurs": http://www.YouTube.com/watch?v=SKqjcPU8xv4 (accessed May 28, 2012).
11. This image is a screenshot from the video "16 Hunnic Turkic Empires": http://www.youtube.com/watch?v=NcbH2sZ_9JA (accessed May 15, 2012).
12. http://www.YouTube.com/watch?v=qkAMWlQ0JPk (accessed May 10, 2012).
13. Tarim mummies were discovered in the Tarim Basin of Xinjiang and date from 1900 BC to AD 200.
14. This image is a screenshot from the video "16 Hunnic Turkic Empires": http://www.youtube.com/watch?v=NcbH2sZ_9JA (accessed May 15, 2012).
15. "Communist China's execution of Uyghur Turks": http://www.YouTube.com/watch?index=5&feature=PlayList&v=asE7CdX1t6c&list=PL5CF9E41A3A98EB11 (accessed May 12, 2012).
16. "Merhum Uyghur Mujahit Berat Haji": http://www.YouTube.com/watch?v=Ni6JYMogKn (accessed May 12, 2012).
17. http://www.YouTube.com/watch?v=-n6GQaR-3Pc (accessed May 11, 2012).
18. http://www.YouTube.com/watch?v=Qhm4N2Cxv24 (accessed May 13, 2012).
19. This image is a screenshot from the video "East Turkestan (Uyghur) genocide by China": http://www.youtube.com/watch?v=du7ndBTbvS0 (accessed on May 15, 2012).
20. See, for example, "Over 45 nuclear tests in East Turkestan by Chinese government. Part 3": http://www.YouTube.com/watch?v=akfpO1cr_kY (accessed May 11, 2012).
21. See, for example, "East Turkestan (uyghur) genocide by China": http://www.YouTube.com/watch?v=du7ndBTbvS0 (accessed May 11, 2012).
22. http://www.YouTube.com/watch?v=P6-YkZml5hY&feature=fvsr (accessed May 12, 2012).
23. "Hasan Mahsum": http://www.YouTube.com/watch?v=UqFZohw7Qak (accessed April 11, 2012); "Islamic Party of Turkistan: Dunya Musulmanlirigha Omumi Mu": http://www.YouTube.com/watch?v=zj2_aH1P4iE (accessed April 11, 2012).
24. This image is a screenshot from the video "Türkistan Islam Partiyisining Bayanati-2010–01": http://www.youtube.com/watch?v=OpSTnpTTFU&feature=related (accessed May 15, 2012).
25. These warnings were issued before the Olympic Games in Beijing in 2008, and because the language used was Uyghur, they could be used as recruitment messages for the youth to join the radical activists.
26. http://www.YouTube.com/watch?v=mbNIX_2rF4M (accessed April 11, 2012).
27. http://www.YouTube.com/watch?v=409ehQ3HY4Q (accessed May 15, 2012).

28. The Great Western Development Plan is a policy adopted by the China to boost its western regions, traditionally less developed than the coastal regions of eastern China. The plan is considered by the Uyghurs as a mere colonization: see, for instance, http://www.nytimes.com/2010/05/29/world/asia/29china.html (accessed May 25, 2012).
29. http://www.youtube.com/watch?v=409ehQ3HY4Q (accessed May 20, 2012).
30. The most notable example is close to Xinjiang. It is the world highest railway track that links Beijing with Tibetan capital Lhasa. Tibetan diaspora raised the issue that the railway serves for transporting Chinese military, not ordinary Tibetans. See "Crossing the line. China's railway to Lhasa" at http://www.savetibet.org/documents/reports/crossing-line-chinas-railway-lhasa-tibet (accessed april 9, 2012).
31. "Hasan Mahsum": http://www.YouTube.com/watch?v=UqFZohw7Qak (accessed April 11, 2012).
32. http://www.YouTube.com/watch?index=5&feature=PlayList&v=asE7CdX1t6c&list=PL5CF9E41A3A98EB11 (accessed May 8, 2012).

REFERENCES

Baranovich, Nimrod. 2003. "From the margins to the centre: the Uyghur challenge in Beijing." *The China Quarterly* 175:726–750.

Becquelin, Nicolas. 2000. "Xinjiang in the nineties." *The China Journal* 44(July):65–90.

Birdsall, William. 2007. *Web 2.0 as a social movement*. Retrieved April 10, 2012 (http://www.webology.org/2007/v4n2/a40.html).

Bovingdon, Gardner. 2002. "The not so silent majority. Uyghur resistance to Han rule in Xinjiang." *Modern China* 28(1):39–78.

Boyd, Danah and Nicole Ellison. 2007. "Social network sites: definition, history, and scholarship." *Journal of Computer–Mediated Communication* 13(1):210–230.

Castells, Manuel. 2007. "Communication, power and counter-power in the network society." *International Journal of Communication* 1:238–266.

Chen, Yu-Wen. 2010. *Who made Uyghurs visible in the international arena? A hyperlink analysis.* (Global Migration and Transnational Politics Working Paper). Fairfax. VA: Center for Global Studies, George Mason University.

Christoffersen, Gaye. 1993. "Xinjiang and the great circle: the impact of transnational forces on Chinese regional economic planning." *The China Quarterly* 133:130–151.

Chung, Chien-peng. 2006. "Confronting terrorism and other evils in China: all quiet on the western front?" *China and Eurasia Forum Quarterly* 4(2):75–87.

Clarke, Michael. 2007. "China's internal security dilemma and the 'Great Western Development': the dynamics of integration, ethnic nationalism and terrorism in Xinjiang." *Asian Studies Review* 31:323–342.

Culpepper, Rucker. 2012. "Nationalist competition on the internet: Uyghur diaspora versus the Chinese state media." *Asian Ethnicity* 13(2):187–203.

Dwyer, Arienne. 2005. *The Xinjiang conflict: Uyghur identity, language, policy, and political discourse.* Washington, DC: East-West Centre.

Fish, Stanley. 1976. "Interpreting the variorum." *Critical Inquiry* 2(3):465–485.

Geukjian, Ohannes. 2007. "The politicization of the environmental issue in Armenia and Nagorno-Karabakh's nationalist movement in the South Caucasus 1985–1991." *Nationalities Papers* 35(2):233–265.

Golder, Scott and Bernardo A. Huberman. 2006. "The structure of collaborative tagging systems." *Journal of Information Science* 32(2):198–208.

Hall, Stuart, ed. 1980. *Culture, media, language* . New York: Routledge.
Hodge, Rodger and Gunther Kress. 1993. *Language as ideology*. London: Routledge.
Hyman, Anthony. 1997. "Turkestan and pan-Turkism revisited." *Central Asian Survey* 16(3):339–351.
Jenkins, Henry. 2007. "Nine propositions towards a cultural theory of YouTube." *Confessions of an Aca-fan: the official weblog of Henry Jenkins*. Retrieved April 10, 2012 (http://henryjenkins.org/2007/05/9_propositions_towards_a_cultu.html).
Kubicek, Paul. 1997. "Regionalism, nationalism and realpolitik in Central Asia." *Europe-Asia Studies* 49(4):637–655.
Lange, Patricia. 2007. "Publicly private and privately public: social networking on YouTube." *Journal of Computer-Mediated Communication* 13(1):361–380.
Lange, Patricia. 2008. "(Mis)conceptions about YouTube." Pp. 87–100 in *Video vortex reader: responses to YouTube*, edited by G. Lovink and S. Niederer. Amsterdam: Institute of Network Cultures.
Lindlof, Thomas R. 2002. "Interpretive community: an approach to media and religion." *Journal of Media and Religion* 1(1):61–74.
Roberts, Sean R. 1998. "The Uyghurs of the Kazakhstan borderlands: migration and the nation." *Nationalities Papers* 26(3):511–530.
Rouse, Roger. 1991. "Mexican migration and the social space of postmodernism." *Diaspora* 1(1):8–24.
Sautman, Barry. 1998. "Preferential policies for ethnic minorities in China: the case of Xinjiang." *Nationalism and Ethnic Politics* 4(1–2):86–118.
Sautman, Barry. 2005. "Preferential policies for ethnic minorities in China: the case of Xinjiang." *Asian Affairs: An American Review* 31(2):87–118.
Shichor, Yitzhak. 2006. "Fact and fiction: a Chinese documentary on Eastern Turkestan terrorism." *China and Eurasia Forum Quarterly* 4(2):89–108.
Vergani, Matteo. 2011. "Folksonomy nel Web, tra utopia e realtà." Pp. 115–139 in *Nuovi media e ricerca empirica. I precorsi metodologici degli Internet Studies*, edited by Simone Tosoni. Milano: Vita e Pensiero.
Vergani, Matteo and Dennis Zuev (2011) Analysis of YouTube Videos Used by Activists in the Uyghur Nationalist Movement: combining quantitative and qualitative methods, *Journal of Contemporary China*, 20(69), pp. 205–229.
Vergani, Matteo and Dennis Zuev. 2012. "Qualitative analysis of the ultranationalist self-presentations in YouTube: comparing visual stories of Uyghur and Chechen separatists." Presentation made at the 40th IIS congress, New Delhi, February 2012.
Wachowich, Nancy and Willow Scobie. 2010. "Uploading selves: Inuit digital storytelling on YouTube." *Etudes/Inuit/Studies* 34(2):81–105.
Waite, Edmund. 2006. "The impact of the state on Islam amongst the Uyghurs: religious knowledge and authority in the Kashgar oases." *Central Asian Survey* 25(3):251–265.
Wenger, Etienne. 1998. *Communities of practice: learning, meaning and identity*. Cambridge: Cambridge University Press.

7 On the Visual Semiotics of Collective Identity in Urban Vernacular Spaces

Timothy Shortell and Jerome Krase

Much of the recent work on global cities examines global capitalism and the function of cities within it. Often neglected in this literature is the role of global flows of people and cultures. The cities identified as global as a result of their position in world economy are often also centers of global migration. Just as the institutions of global capitalism have changed urban places and spaces, so too has the mass movement of people from one local community to another. In the vernacular spaces of urban neighborhoods, the effects of migration are easily seen.

Visual analysis of social space in urban vernacular neighborhoods reveals much about the phenomenological nature of globalization. An examination of the quotidian rhythms of urban life is as important to our understanding of the nature of globalization as the analysis of large-scale social forces and institutions. The everyday lives of urban dwellers are shaped by forces outside of their control, certainly, but people assert agency in a multitude of ways. We can see this in their practices when they perform identity, making the social spaces they occupy like "home."

When we move through urban spaces that we haven't seen before, such as residential neighborhoods or commercial thoroughfares, we become natural ethnographers, or tourists, attempting to decode and decipher the visual clues and cues that shout and whisper to us along the way. Many of us will ask ourselves whether this is a safe or a dangerous place. "Am I welcome here or should I leave before it is too late?" "What kind of neighborhood is it?" "Are the people who live and work here rich or poor?" "What is their race, ethnicity, or religion, and how (or why) does it matter?" Some things are easy to translate from the languages of the street, such as what is for sale. Legitimate merchants make it obvious that they are seeking customers with intentional signs that compete for attention, but for the sale of illicit goods like drugs and sex, the signs vendors give off are subtler. Yet even for illicit goods and services it is clear that for the knowledgeable customer they are in "plain view." This reading of the vernacular "street signs," so to speak, is not merely an aesthetic exercise. What we see makes a difference in how we respond to the places and the people we encounter in our increasingly complex and changing urban surroundings.

The creation and maintenance of society is dependent on our senses, and vision is perhaps the most powerful of them. Beyond neonatality, our first experiences of life as "social life" are mostly visible ones, as when we encounter, recognize, and differentiate between people. It has also been consistently argued in the social sciences that face-to-face—therefore also eye-to-eye—interactions in primary groups are the building blocks of subsequent social life.

The everyday practices of ordinary people who have been brought together by globalization are examples of what Robertson (1997) meant when he coined the term "glocalization." According to Osterhammel and Petersson (2005:7), "Robertson recognized that homogenizing and universalizing forces of globalization do not obliterate the heterogeneity and particularity of local forces as much as their interaction creates varying degrees of hybrid culture." People perform their collective identity by making use of the materials of this hybrid. Their practices and symbols might not be identical to those of their culture of origin, but they are no less meaningful as a result.

In this chapter, we argue that sociological analysis of visual data is necessary to understand how urban vernacular neighborhoods are changing as a result of globalization. Visual data reveals the constructed and dynamic nature of social meanings and the ways in which urban space is both the context of and product of race, ethnic, class and cultural transformations. We will briefly document some of these transformations with photographs of vernacular neighborhoods in Brooklyn, Cape Town, Gothenburg, Los Angeles, Manchester, Paris, and Philadelphia, along with additional descriptions of similar scenes from the more than 9,000 photographs in our online archive of urban neighborhoods in global cities.[1] Urban spaces are filled with signs of collective identity and, often, inter-group competition. In the physical environment, architectural details, commercial signs, and graffiti, among other things, signify the flows of people and culture. So too do social practices, such as commercial transactions, collective action, socializing, and commuting, in the public spaces of vernacular ethnic neighborhoods. Our analysis, based on the images shown here and hundreds of others, reveals what glocalization looks like and demonstrates the connection between social space and collective identity. We bring together insights from semiotics and symbolic interactionism to explain some of the ways people perform identity in multicultural social spaces.

IMMIGRANT IDENTITY IN URBAN COMMUNITIES

Sociologists arguing for the primacy of class in explaining the nature of urban environments, such as Gans (1991), have tended to regard ethnicity as one more variable affecting an individual's life chances, significant to the extent of these effects. Waters (1990:194) makes a similar point about the connection between ethnicity and prejudice, noting,

> the degree of discrimination against white European immigrants and their children never matched the systematic, legal and official discrimination and violence experienced by blacks, Hispanics, and Asians in America. The fact that whites of European ancestry today can enjoy an ethnicity that gives them options and brings them enjoyment with little or not social cost is no small accomplishment. But does it mean that in time we shall have a pluralist society with symbolic ethnicity for all Americans?

But this view diminishes the phenomenological component of identity in favor of the structural by calling some kinds of ethnicity "symbolic"—as if to imply merely so—because some groups have not been subjected to the same kind and extent of prejudice in their host societies. But because ethnicity involves the meaning-making processes of culture, whether attachment to an identity is voluntary or ascribed, all ethnicity is both symbolic and material simultaneously. It is neither more nor less a factor in urban life for being such.

Many immigrants in global cities today are experiencing some of the kinds of discrimination that historically attached to peoples of color in the U.S. This is particularly true for immigrants from the Islamic world in many European nations (and the U.S.). Even in the absence of this kind of "Jim Crow" discrimination, the social spaces inhabited by immigrants tend to be marked with diminished status, a stigma or an "otherness." The effects on the day-to-day life of urban dwellers in these communities are substantial and cannot be reduced to a differential set of life chances.

A related argument concerns the "authenticity" of ethnic identity. Krase (2008:7) has argued that authenticity is a matter of agency, of the power of ordinary people to make meaning through their cultural practices; he notes,

> In my view, as opposed to that of a more demographically oriented scholars such as Richard D. Alba, there can be no "Twilight of Ethnicity," if we take twilight to mean only "sunset" and not also "sunrise." Furthermore, to say that something is not "authentic" is to deny the agency (read: possibility of authorship) of the individual or the group making the claim to it.

We argue here that spatial semiotics reveals how ordinary people have the ability to create meaning by changing the appearances of places and spaces. It doesn't matter that "glocal" cultural practices differ from their geographic or historical models. Rather, what matters is that ordinary people use these practices to perform their collective identity.

In his analysis of Los Angeles, Davis (2001:15) argued that "Hispanic/Latino performance can be viewed as practice as well as representation." There is a simple formula for the process by which agency transforms practice into representation in vernacular landscapes: ethnic groups simply going about their daily business present themselves to the observer. The observer then re-presents their performances in descriptions that in turn become representation—in some cases, stereotypes, and in others, commodified icons of otherness. Beauregard and Haila (2000:23) note that despite the increased spatial complexity of late twentieth-century urbanism, a distinctly "postmodern" city has not displaced the modern one and is just as "legible" as its precursor. Lofland (1985) might add that cities have always been changing in response to the entrance of "strangers." The difference today is primarily the rapidity and variety of that change that follow a different logic of location.

Fritzsche (1996), Lynch (1960), and King (1996) speak of cities as "texts" to be read and, we argue here, ordinary streetscapes are important yet often ignored parts of those texts. Zukin (1996:44) adds, "Visual artifacts of material culture and political economy thus reinforce—or comment on—social structure. By making social rules 'legible' they represent the city."

At the street level, the symbolic capital of immigrant and ethnic entrepreneurs is reflected in the changing appearance, and therefore also the meanings, of commercial streetscapes. For Bourdieu (1977), symbolic capital serves to obscure relations of domination and the reproduction of the class hierarchy. Some of the "hidden" reproductions of power as re-presentations of ethnic or class identity, however, are "in plain view" as people follow the course of their everyday lives. Bourdieu's notion of the "habitus," or practices that produce social regularities—which we argue become visible signs of identity—is also helpful in this regard.

SPATIAL SEMIOTICS

Simmel (1924[1908]) established the central role of the visible in theorizing about the complex and constantly changing metropolis. He noted the extent to which modern cities made social differences part of the visual landscape, the way that cities looked. Urban dwellers, he argued, changed the ways that they related to the built environment and to each other as a result. Combining Simmel and other seminal urban theorists, such as Lefebvre (1991), Lofland (1985, 2003), and Jackson (1984), Krase and colleagues (Krase 2002, 2003; Krase and Hum 2007; Krase and Shortell 2008; Shortell and Krase 2009) have demonstrated that ordinary people change the meaning of spaces and places by changing their appearance, through their activities and by their presence.

For Lefebvre (1991:75–76), the visual was central to the production and reproduction of social space of any scale:

Thus space is undoubtedly produced even when the scale is not that of major highways, airports or public works. A further important aspect of spaces of this kind is their increasingly pronounced visual character. They are made with the visible in mind; the visibility of people and things, of spaces and of whatever is contained by them. The predominance of visualization (more important than "spectacularization", which is in any case subsumed by it) serves to conceal repetitiveness. People look, and take sight, take seeing, for life itself. We build on the basis of papers and plans. We buy on the basis of images. Sight and seeing which the Western tradition once epitomized intelligibility, have turned into a trap: the means whereby, in social spaces, diversity may be simulated and a travesty of enlightenment and intelligibility ensconced under the sign of transparency.

Harvey (1989, 2006) argued that the powerful reproduce and enhance their power by controlling public space. Through appropriation and domination, the powerful differentiate public space. The lives of ordinary urban dwellers take place in this context. Although disadvantaged in their struggles with the powerful, ordinary urban dwellers are not powerless. They contest and sometimes subvert domination by using public space for their own ends, sometimes through collective action and sometimes by "unofficially" being in the space.

Our visual semiotic work on globalization in urban neighborhoods owes much to the many contributions to urban studies of Lofland and her application of symbolic interactionism. She noted that her fellow interactionists have contributed greatly to our understanding of urban worlds by demonstrating how people communicate with each other via the built environment. One of the ways this is done, for example, is by the common practice of seeing settlements as symbols (Lofland 2003:938–939; see also 1985, 1998). In urban worlds, individuals and groups interact with each other through visual images that in turn influence what people encounter on the streets. The meanings of what they see, however, come from a different source—meanings of symbols are learned through socialization, via primary and secondary affiliations. Lofland (1985:22) also argued that "city life was made possible by an 'ordering' of the urban populace in terms of appearance and spatial location such that those within the city could know a great deal about one another by simply looking." This need to create order is especially true today in the increasingly transnational, home territories that modern migrants seek to create and modify.

Lofland (1991) was puzzled by the apparent absence of major contributions by interactionists to the field of urban sociology despite their having equally deep roots in the Chicago School. It was odder given the broad range of topics to which the "urban" label could be attached. She argued that if a more "analytically rigorous" definition of the field was used, clearly the work of Anselm Strauss was crucial. For visual sociologists Strauss' discussions on urban imagery are also instructive.

In approaching the city—the urban settlement form—as a research topic, Strauss eschewed the standard questions of the time. He did not ask whether urban people are necessarily alienated and estranged from community or whether cohesive and integrated neighborhoods could be found. Nor did he ask how the typical division of American cities into "natural areas" came about or how population movements affected land values and patterns of land use. Instead, Strauss approached the city *as* a research topic in a manner that is quintessentially interactionist: he asked about *meaning*. He asked about "what Americans think and have thought of their cities" (Strauss 1961:viii); he asked about the "symbolic representations of the urban milieu" (Wohl and Strauss 1958). He asked how Americans define the urban situation, how they interpret the kaleidoscope of sights and sounds and smells that is the urban environment. He asked, in sum, about urban imagery (Lofland 1991:207).

From the foundation of this interactionist approach, we have developed an analytical perspective on spatial semiotics that pays close attention to the nature of signs as such. Social interaction is based on shared meanings, constructed through various forms of individual and group behavior in the social spaces of urban communities. But social meanings cannot be taken for granted; the process of communication—of creating signs and interpreting them—is complex and imperfect. For urban dwellers, literacy of urban semiotics is for the most part implicit and, like much human communication, prone to stereotypes, and like all forms of social knowledge, structured by various status hierarchies.

Vernacular landscapes are the interpretive context of the signs of collective identity of interest in the present research. Signs have meanings that relate to the patterns and places of urban life. These give sensibility to the "visual impressions" that Simmel (1924[1908]) so thoughtfully observed. The vernacular landscape is both the built and social environments, what Gottdiener (1994) called "settlement space." Lived experience in urban communities, as well as media sources about urban culture, is used to make sense of the signs of collective identity.

The semiotic medium for collective identity messages in urban space is multimodal, including linguistic, spatial, and visual signs. There are codes for appearance, for example, and nominal codes. Physical characteristics, such as skin tone, hairstyles, and physiognomy, are particularly important signs of ethnic identity. There are visual codes that relate colors to identity and codes for alphabets as physical signs of geography. Distinctive cultural practices are also a code; perhaps the most common of these relate to food and dress.

The markers of collective identity are in a constant state of tension among competing possible interpretations, especially between in- and out-groups. That the producer of a sign and interpreters of it may have different meanings in mind is an important and often overlooked aspect of urban social space.

Jakobson (1960, 1972) identified four functions of signs that can help us interpret visual representations of collective identity in urban neighborhoods: *expressive*, *conative*, *poetic*, and *phatic*. Jakobson was referring to the operation of language, but the functionality he describes may be applied to spatial and visual signs also. In doing so, we are extending Jakobson's semiotics in a manner that may be useful for urban researchers.

Expressive signs give the subject a voice; they are an important component of social agency. According to Jakobson, these signs reveal the affective state of their creator, such as their feelings about their home culture. In the context of urban neighborhoods, people create expressive signs in the course of their everyday practices when they enact rituals of identity. Among the most visible of these practices is the use of flags, national colors, or place names to proclaim origins.

In Jakobson's view, conative signs attempt to influence others' behavior. In language, the use of vocative case or imperative mood signals the conative function. Conative signs highlight the relationship between the addresser and addressee, and place an obligation on the latter. Markers of exclusion are one important type of conative sign in the urban vernacular landscape. Graffiti can be said, in some instances, to be of this type. So too can the uses of different alphabets (relative to the majority language—for some groups, some kinds of writing will be recognized as such but remain inscrutable). To the extent that social space is contested, conative signs are common; they call attention to group boundaries, marking the space between the in-group and out-groups.

Poetic signs express style; they represent the aesthetic dimension of communication of identity. Urban styles, like hipster or hip-hop, clearly have a poetic component. The practices that make up such styles are routinely commodified or used as stereotypical icons of urban life. But in their ubiquitous quotidian forms, poetic signs mark urban space as "belonging to" a particular group, often in subtle ways directed at those "in the know."[2]

Jakobson describes phatic signs as those that are oriented toward contact (Hawkes 2003). In language, this includes, among other things, phrases which facilitate continued communication rather than are strictly denotative. For example, in conversations questions such as "You know what I mean?" function as tests of the connection between addresser and addressee. Applied to visual codes, phatic signs are those that serve as an inducement to social interaction. They work not through denotation primarily, but by confirming the connection between individuals, and between people and places.

In the end, visual signs that facilitate social relations, like these linguistic customs, might be the most common signifier of the urban vernacular. Phatic signs are artifacts of ordinary social interaction. They are the indicators that we are at home in our neighborhood. Phatic signs express that a social space belongs to us (at least for a time), that our cultural practices are acceptable there. Through phatic signs, cultural strangers can assert their agency in the social spaces of the host country.

The main distinction between expressive and phatic signs is that production of the latter is not primarily about advertising identity. Goffman (1959:2, emphasis in the original) makes a similar distinction in contrasting two kinds of sign activity, "the expression that he *gives*, and the expression he *gives off*." The former is intentional communication, often with verbal behavior, and the latter involves the interpretations observers make of a wide range of attributes and behaviors performed in public. People generally attribute motives when interpreting expressive signs, but phatic signs are material manifestations of in-group normative practices in local spaces. Outsiders can read them as indicators of collective identity, even if that was not the intention of those whose practices enacted the signs. Urban dwellers of vernacular landscapes are generally literate with regard to signs of collective identity, but attribution errors are common. We tend to see other groups' identity performances as directed toward us even when they are not.

METHOD AND DATA

Visual sociologists generally use one or more of three different kinds of images in research. Some researchers have research subjects produce images. Other researchers use found or preexisting images as data. In the present research, we employ the third kind, researcher-produced images (Warren and Karner 2005; Pauwels 2008). We use a visual method called the photographic survey (Krase and Shortell 2007, 2008, 2009; Shortell and Krase 2009) to collect data. The photographic survey is a technique for taking images of urban neighborhoods in order to record visual information at a particular place and time. Photographs are taken as the researcher travels through a neighborhood systematically photographing public spaces, without regard to particular content or aesthetics.

The photographic survey records both the physical and social streetscapes. It is important for data collection not to be determined by the researcher's attention—that is, the researcher must not photograph only that which seems, at that moment, to be of interest. The photographic survey is designed to collect images in which the social content might not be immediately noticed. This overcomes the most important shortcoming of most studies using researcher-produced data: sampling bias. Like the ethnographer collecting observational data, this method produces a lot of visual information in the field, the significance of which may be known only later, upon reflection.

Multiple trips are required to adequately cover a particular neighborhood. Because social life in urban neighborhoods is dynamic, varying by time of day, day of week, and week of year, as well as year to year, more trips to photograph a neighborhood results in data with greater validity. It becomes possible to see patterns when the collection of images extends beyond the boundaries of particular urban cycles. The end result of data collection using the photographic survey of a particular place is usually hundreds of photographs.

Following Harper (1988), we adhere to a visual method in which the photographs are data, records of the structuring of social life, not mere ornaments to illustrate sociological concepts. We regard the photographic survey as a kind of visual ethnography where the data, the images, serve more than important documentary or purely descriptive purposes but as data (re-presentations) to be analyzed within a theoretical framework. We apply a framework that is based on structural semiotics and symbolic interactionism, but other researchers could use our data in other ways. Although it can be argued that, as is true of all comprehensive data collection processes, the photographic survey produces images that are insignificant for the narrower intended purpose of a particular analysis, the extraneous data, as valid photo documentation, can be useful in other analyses. To study spatial semiotics of urban vernacular landscapes, researchers must use visual data of the physical and social spaces of urban neighborhoods. Just as with non-visual sociologists, researchers have to be attentive to the ways in which data collection shapes the possibilities of analysis.

All image data for this project are available online; the photo archive contains more than 9,000 images from more than forty global cities (see note 1). Images for the present analysis include neighborhoods in Berlin, Brooklyn, Cape Town, Frankfurt, Gothenburg, Lisbon, Los Angeles, Manchester, Milan, New Britain, New York, Oslo, Paris, Philadelphia, St. Petersburg, and Stockholm.

The photographic survey addresses one of the biggest problems with ethnographic research, the lack of generalizability that comes from non-probability sampling, by devising a practical data collection strategy that is not limited by the researcher's attention. This solution generates a large number of images, which creates a burden for analysis, but that is a trade-off we believe is worthwhile; the logic is inductive, like grounded theory approaches—begin with a lot of data and build from there. The photographs produced by this method are often ambiguous and the technique is only as good as the application of a particular theoretical framework allows, but that is true of other ethnographic methods as well. We believe this method is particularly well suited for the study of social interaction, in part because it is unobtrusive and allows us to see the context in which interactions take place. In our view, there is no better way to study the urban vernacular landscape.

SCENES FROM URBAN VERNACULAR SPACES IN GLOBAL CITIES

Figure 7.1 shows some instances of phatic signs of collective identity. In image (a), clothing functions as a phatic sign; the style of dress of these

Orthodox Jewish women on Coney Island Avenue is a well-known indicator of religious/ethnic identity in Brooklyn. Businesses that provide goods and services of interest to immigrant groups, such as the communications services advertised in the window of this storefront in Philadelphia, as shown in image (b), represent another kind of phatic sign. Not only the language of the signage, but also the particular array of cards and rates signals the potential customers of this business. We can read who the local market is by observing such emphasis. Image (c) shows the "Little Aladdin" restaurant on High Street in Manchester, in the Northern Quarter. The area is known as the "hipster" neighborhood, which might suggest that it offers Indian food for non-Indians. But Manchester has a sizable South Asian population nearby, and the window promises "it's just like home cooking"—a message seemingly oriented toward those for whom Indian dishes would be their home cuisine. Image (d) shows how cultural practices function as a phatic sign. This a message board in a public plaza in Angered Centrum, an ethnic suburb of Gothenburg, displays a variety of

Figure 7.1 Phatic signs of collective identity. Clockwise from upper left: (a) Orthodox Jewish women on Coney Island Avenue in Brooklyn—photograph by Timothy Shortell, 2010; (b) communication services advertised on a window in Philadelphia—photograph by Jerome Krase, 2003; (c) "Little Aladdin" on High Street in Manchester—photograph by Timothy Shortell, 2009; and (d) message board in a public plaza in Angered Centrum, an ethnic suburb of Gothenburg—photograph by Timothy Shortell, 2010.

announcements for events. Unlike advertisements in other media, which might reflect the homogenizing effects of globalization, message boards like this don't have much reach; they are for local consumption only.

Two examples of clothing store windows from our photographic archive demonstrate the visual similarity of the kind of phatic signs shown in Figure 7.1, image (a).[3] The window of the "Asian Cloth House" on Tovengt Street in Olso advertises fashions directed at a particular ethnic minority. The same can be seen in a window of a shop on Rue du Faubourg Saint Denis in the tenth arrondissement of Paris. The style of dress reflects a distinct cultural marker; although anyone is free to shop in these places, the presence of the window displays—as well as people wearing these fashions on the street—marks the territory as a "South Asian neighborhood."

Money transfer and telecommunications services, such as those found on the commercial streets in Belleville, Paris, are common where there are significant numbers of immigrants in global cities on every continent. This part of Belleville is pan-Asian, and the services reflect this. A nearby CD and DVD store sells entertainment and other specialty items from East Asia. Similar businesses can be seen in the Pakistani neighborhood along Coney Island Avenue in Brooklyn, also use alphabets on signage as an indicator of who lives here.

Food is a ubiquitous phatic sign. Food choices reflect cultural practices. The logic of commerce results in food shops stocking and advertising products that are in demand by the local community. For example, there are many Chinese markets in Sunset Park, Brooklyn, where the large Chinese community in the neighborhood shops; the concentration of these businesses ascribes the neighborhood as Brooklyn's "Chinatown." Non-Chinese residents, as well as outsiders, tend to see the display of these foods as evidence that Sunset Park is a Chinese neighborhood. South Asian stores along Coney Island Avenue (mostly Indian and Pakistani groceries) function in the same way. The concentration of these shops effectively designates this part of Coney Island Avenue as "Little Pakistan" in Brooklyn. The accuracy of the label is less important than the fact of the visual labeling process. For example, Sunset Park's Chinese population was in the majority before it was called a Chinatown—which suggests that visibility is not a purely demographic phenomenon. It wasn't until the commercial street reflected the group (and the group's growing economic clout and social status) that it earned the label. Similarly, Pakistanis are a distinct minority on Little Pakistan where Bangladeshis and other South Asians dominate the residential scene.

Ethnic cuisine is, of course, global and is not necessarily a reliable indicator of a neighborhood's population. The "authenticity" of such cuisine is often a marketing tactic, not a historical argument. Nonetheless, some ethnic restaurants are manifestations of glocalization. For example, the façade of "Istanbul Grill" on Potsdamer straße and Bulöwstraße in Berlin advertises *döner*, the rotary grilling style of Turkish cuisine. Germany hosts a significant Turkish population, making these restaurants more common in contemporary German cities. Although *döner* is becoming a popular "street food" in global cities with Muslim populations, such restaurants still signify immigrant spaces.

Phatic signs also signify class identity. Consider, for example, an informal flower market in St. Petersburg near a major metro station. The vendors of the flowers are identified as belonging to the bottom of the class hierarchy by the fact that they do this kind of labor, as well as their appearance. Many shoppers on the street tend to ignore these workers because of the stigma associated with their lack of status. In contrast, in Prospect Heights, Brooklyn one can see many blocks of brownstones on quiet residential streets. The neighborhood is a hot zone of gentrification, and the social space of blocks like this are assigned to the upper middle class.

In Figure 7.2, expressive signs of collective identity are demonstrated. Image (a) shows the minaret of a mosque in the gentrifying multicultural Bo Kaap section of Cape Town. The architecture is distinctive and makes an effective icon of religious identity. Religious buildings often function in this way. The visibility of this architecture—that is, the extent to which it is generally noticed—can become the focus of social conflict, as is the case with hostility toward the building of large mosques by Muslim communities in many global cities at present. Image (b) shows a banner hanging above the street market in Belleville, Paris. The banner exclaims "Belleville solidarity," a reference to the anti-gentrification movement that many of the immigrant groups in Paris' traditional immigrant neighborhood had participated in, which had stopped some of the planned redevelopment, saving some of the existing low-cost housing in the area. In image (c), a flag mural in Williamsburg, Brooklyn, is shown. Flags as icons of national origin are commonplace expressive signs of ethnic identity and are often interpreted, as are gang graffiti, as ways of claiming contested territory as ones own (combining expressive and conative functions). In this case, the gentrification of the neighborhood has put pressure on the Puerto Rican and Dominican working-class neighborhood; the presence of Latino stores and other local businesses between Bedford Street and the waterfront, south of Grand Street, is diminishing. Flags, and flag colors, are incorporated into commercial signage, residential facades, as well as street art such as this. Image (d) provides an example of expressive signs in a commodified form, the "ethnic theme park." Krase (1997:105) explains the ethnic theme park as a place where the experience of the ethnic other is for sale, particularly to tourists. Monuments to the group's ethnic "heritage" and "anthropological gardens" where visitors can see glimpses of the "good old days" are common signifiers. The image shows a scene from the iconic Chinatown in Los Angeles. The "traditional architecture of the "Hop Louie" restaurant, and the dozens of similar buildings, attracts thousands of tourists yearning for a glimpse of the "authentic Chinese" way of life but who instead consume its simulation.

Figure 7.2 Expressive signs of collective identity. Clockwise from upper left: (a) minaret of a mosque in Cape Town—photograph by Jerome Krase, 2000; (b) "Belleville Solidaire" banner at public market in Paris—photograph by Timothy Shortell, 2007; (c) Puerto Rican flag mural, Williamsburg, Brooklyn—photograph by Jerome Krase, 2010; and (d) ethnic theme park in Los Angeles—photograph by Jerome Krase, 2001.

Additional instances of these expressive signs can be seen in our photographic archive. A storefront mosque in La Goutte d'Or in Paris uses the traditional colors of Islam, green and white, as well as the Arabic alphabet to announce itself. Both the signage and the function of the building are expressive signs. There are often worshippers outside, further marking the territory as home to the North African Muslims who live in the surrounding blocks. A similar example can be seen in an image from Frankfurt, where the signage for a mosque blends in with the ethnically sensitive commercial signage on the block, including a *döner* shop.

There are many examples of collective action serving as an expressive sign of collective identity in our archive. For example, we photographed a pro-democracy protest in Piccadilly Gardens by Manchester's local Iranian community. The participants and the protest signs and flags they carried were clearly

On the Visual Semiotics of Collective Identity 121

expressing a connection to Iran, temporarily marking the park as their space. Among our images of Paris' immigrant neighborhoods is a photograph of a march for the rights of *personnes sans papiers* (undocumented immigrants) in Paris. The march appeared to be headed toward Place de la République, a traditional space for populist collective action, such as the marches by North African youth in the mid-1980s (Derderian 2004). A photograph shows the march turning onto Boulevard de la Chapelle from La Goutte d'Or, a neighborhood of North African and West African immigrants.

Explicit performances of culture also function as expressive signs. Examples of such performances, both the spectacular and the quotidian, can be found in our archive. A photograph from New York City, for example, shows a scene from the annual Turkish Day parade in Manhattan that is one of many examples of how organized ethnic and other types of groups attempt to positively manage their identities via public performances. Another image shows a pair of Andean musicians performing for tips (and selling CDs) in Stockholm; buskers like this are common in global cities where migrant newcomers compete with locals for both attention and making a living. Their native costumes are generally as noticeable as is their different music and dancing.

A variation on the flag theme is displayed in photograph from New Britain. The fascia of the "Polish-American Congress" office uses the coat of arms and red and white of the Polish flag, shaped like the state of Connecticut, to announce its ethnic identity. By incorporating the Polish national colors in the shape of Connecticut the group can also be seen as making a claim of "Americanness" or at least hyphenated/hybrid identity.

The plethora of expressive signs demonstrates the variety of motives involved in the performance of identity. In urban spaces where immigrants are, temporarily (such as transit hubs) or more permanently (such as residential neighborhoods), the motive is frequently commercial—economic push and pull factors remain highly significant in global population flows. Some of the behaviors are normative, taking into account the possibilities of the host communities; baseball caps and soccer/football jerseys, for example, are popular urban fashions. In New York City, the frequent ethnic celebrations provide opportunities for expressive behavior. In Paris, to give another example, some spaces (such as Belleville or La Goutte d'Or) are known to be historically immigrant places; residents can perform their identity there in ways that are riskier in other districts. Some signs are employed because of convenience or expedience; flags and national colors are both widely available and especially effective at communicating place information. Finally, some signs are habitual; their producers know they are communicating identity through tradition even if they don't know the full historical genealogy of a particular practice.

Figure 7.3 shows conative and poetic signs of collective identity. In image (a), the menus for two restaurants in Manchester's Chinatown are displayed. The restaurant on the left has a bilingual menu. But at the restaurant on the right, the menu is in Chinese only (except for the name of the restaurant, in

English). This would be an invitation to Chinese speakers, but a significant disincentive for those lacking literacy in the language. The iconic markers of urban poverty in the Global South, the shantytowns, slums, and *favelas* are usually interpreted as signs of danger to outsiders. In image (b), this township outside of Cape Town could almost stand as an easily recognizable sign for urban poverty throughout the developing world. As noted by Gehl (2010), South Africa has tried, with its "dignified places programme" to create quality urban space for the poor to counter this prevalent symbol. Poetic signs of "hipster" identity are ubiquitous; image (c) displays a poster calling for solidarity with ten people accused of starting a riot in the detention center in Vincennes, which the poster notes is the largest prison for undocumented immigrants in France. The imagery of the poster, as well as the language, recalls the ambience of the May 1968 revolt; the poster explains, "Being in solidarity with the defendants of the fire in Vincennes is solidarity with all those who in one way or another are rebelling against a world where millions

Figure 7.3 Conative and poetic signs of collective identity. Clockwise from upper left: (a) Chinese restaurant menus in Manchester—photograph by Timothy Shortell, 2009; (b) township outside of Cape Town—photograph by Jerome Krase, 2000; (c) political poster calling for solidarity with undocumented immigrants—photograph by Timothy Shortell, 2010; and (d) street art and graffiti in Lisbon—photograph by Timothy Shortell, 2009.

of lives are hanging from bits of paper." Image (d) shows street art in the Barrio Alto section of Lisbon, another poetic sign. This is an area of clubs and bars, and attracts a lot of tourist traffic for its cafes and *fado* performances. The street life is very much youth oriented, and the graffiti and street art forms a backdrop to the shopping and dining.

Conative and poetic signs can be seen in many images in our archive. For example, the presence of Polish on the fascia and awning of a pharmacy acts as an invitation to the longstanding Polish community in Greenpoint, Brooklyn, but not to others who don't understand Polish. The pharmacy is a franchise in a popular chain, but this one is customized for the local community. The monolingual Hebrew and Yiddish signs on some businesses and organizations in Midwood, Brooklyn, have the same effect. This kind of conative sign both welcomes the in-group (those who understand the language) and repels the out-groups (those who don't).

Gentrification generates an abundance of conative signs of class identity, as gentrifiers mark the territory as having acquired new status. Part of that status is in exclusivity, that not everyone is welcome, or "deserves" to be, in the space. In the Lisbon album in our online archive, one can see a busy shopping block in the heart of the city's gentrified downtown, Praça Dom Pedro IV. Here local history, near the site of the original market by the port, mixes with globalized luxury brands. Similarly, a streetscape along Vasagaten in Gothenburg from our archive displays conative signs of class. Although not quite as well known as its more gentrified and touristy neighbor, the adjacent Haga district, this area is lined trendy restaurants and shops, basking in the status of *Göteborgs Universitet*. This is not a place for the faint of wallet.

Subculture identities, such as hipster or hip-hop, often embed poetic signs in the social space of vernacular neighborhoods. In several of our city albums in our online archive one can observe markers of American hip-hop, for example, including Paris, Manchester, Milan, and Berlin. Fashion is one important way that hip-hop identity is signaled. This is different from the phatic signs embedded by clothing practices in that it results from actions that are more deliberately focused on style—and are often motivated more by individual expression than normative practices.[4]

All things upscale are signs of upper middle class status, including the "ironic" use of working-class features, such as business names and façades. Our archive includes numerous examples from gentrifying "hot spots" in Brooklyn. This is a powerful type of poetic sign because it requires the viewer to know that the working class reference is not meant to indicate actual class status, but instead, an urban style.

DISCUSSION

Vernacular landscapes are everywhere and changing as a result of the array of forces associated with globalization. The result is an urban social space layered by signs of collective identity of peoples from distant places. As people move from place to place, they take signs of their home culture and embed these through their presence and their cultural practices in social spaces of their new neighborhood in the host nation. Immigrants generally lack the power to recreate the valued spaces of their home cultures, but their day-to-day lives are full of expressive, conative, poetic, and phatic signs of their collective identity. The urban vernacular landscape has traces of a most diverse array of ethnic, class, religious, and social influences, as seen in the material and interactional forms associated with their home culture—everything from architectural styles and street art to fashion and food preferences.

Glocalization is a hybrid, and manifests itself in visible signs in vernacular neighborhoods. The goods and services desired by locals in immigrant neighborhoods ascribe the place as ethnic, and often, working-class or poor, by virtue of the concentration of physical markers of identity and the spatial distribution of people. It is a tangible proof of the agency of non-elite urban dwellers that they can define the meaning of their social spaces, to some extent, by their presence and their efforts to make their places "home." Our photographic data reveals, over and over again, evidence of this in vernacular neighborhoods in global cities.

Through everyday behaviors, people perform their ethnic, class, religious, gender, and subcultural (affinity group) identity for others—those who share social spaces with them, as residents or visitors. Working, commuting, socializing, shopping, walking, eating, sitting, and even playing football, among many other things, are ways that urban dwellers communicate their identity by being in social spaces in vernacular landscapes. These identity performances are not necessarily motivated by the desire to communicate identity. Rather, they embed phatic signs in urban social space. They are the most ubiquitous signs of collective identity that we have observed.

Phatic signs are crucial to everyday social life in urban communities because it is often impossible to know personally everyone in one's in-groups; cities are, after all, places of strangers. The cultural practices of dress and interpersonal interaction, such as maintenance of personal space or gestures, help to establish trust among people who share culture but may not be personally known to one another. To those in the in-group, these signs communicate that one "knows the rules," thereby facilitating interaction. But to others, these signs also communicate identity.

Expressive signs are another important feature of the vernacular landscape. Through both individual acts and collective behavior, expressive signs of identity become commonplace in urban communities. Whether or not these are the same acts and behaviors that one would observe in the home countries does not matter in this regard. Glocalization is a hybrid, a mixture of home and host milieux. What is most significant about expressive signs is that they are an important and effective manifestation of agency among the relatively powerless, as immigrants in global cities generally are.

Of course, in terms of organized social movements, the creation of expressive signs has had an enormous impact on politics and policy beyond urban space as well.

Conative signs are common in contested spaces. The practices that Lofland (1998) describes regarding the temporary occupation of—taking "possession of"—public spaces represent conative signs. Similarly, the social practices DeSena (2005) studied in Greenpoint, Brooklyn, constitute conative signs because they are the local practices by which groups "protect their turf." Conative signs are also ubiquitous in neighborhoods where one group has achieved demographic dominance because the commercial signage and local businesses (food, clothing, and entertainment, especially) are oriented toward the dominant group, which makes members of outgroups feel less at home.

Poetic signs may seem at first glance more frivolous to the urban researcher, less critical to the marking of urban public space. It is true that they are somewhat harder to find than phatic or expressive signs. But collective identity includes both primary and secondary affiliations and poetic signs play a significant role in the practices of the latter. Urban subcultures, as well as political parties and social movements, generate poetic signs by their use of public space.

These signs of collective identity—phatic, expressive, conative, and poetic—ascribe some urban social spaces as "immigrant spaces." These signs are evidence of glocalization. They reveal, as our spatial semiotic analysis has shown, that vernacular landscapes are dynamic, reflecting the agency of ordinary urban dwellers brought together by population flows. Some of these vernacular places are contested spaces, reflecting competition for housing or employment, among other things. Some of these places are what Anderson (2011) calls a "cosmopolitan canopy."

Despite the differences in appearance of urban space in global cities, and the different immigrant groups and ethnic majorities that occupy them, our spatial semiotic analysis demonstrates some of the common features of "immigrant spaces." Globalization is often said to have a homogenizing effect (Ritzer 2003). But the similarity we describe here is not the result of homogeneous culture. Rather, it is glocal, truly a mixture of home and host cultures. The similarity is structural; it relates to the ways that collective identity gets embedded in social space. This is what makes 'immigrant spaces' in cities such as Berlin, Brooklyn, Gothenburg, Paris, Rome, or St. Petersburg recognizable as such.

As previously discussed, we employ the comprehensive photographic survey method in order to collect visual data (images) in situations where social contents might not be easily apprehended in order to avoid sampling bias. This method produces a lot of visual information in the field, the significance of which may be known only later, upon reflection. Most importantly, the method is an *attitude* toward the images that are produced that serve more than documentary or descriptive purposes but as data (re-presentations) to be analyzed within some theoretical framework; in this research, our concern is the visual semiotics of vernacular landscapes. Other researchers, using the same images, might find different insights by applying a different framework.

Signs of collective identity get embedded in the social spaces of urban vernacular landscapes through intentional and habitual performances, both individual and collective behavior, in obvious and subtle ways. As interactionists such as Lofland have repeatedly shown in so many different urban contexts, we are interpretable objects for each other, both in our appearance and our behavior. To see collective identity in social space, it is not enough to ask urban dwellers about their identity or their social behavior—or to argue about whether or not it is authentic. Researchers must investigate the many and sometimes competing interpretations of the signs of collective identity, both quotidian and spectacular, by close inspection of the visual and spatial semiotics of the urban vernacular. Urban dwellers communicate with each other, whether or not they intend to or are fully aware of it, by changing the way social spaces look through patterns of cultural practices and by their presence—which combine in complex ways to constitute "visibility." Our analysis explains some of the ways that this communication is accomplished.

Because the signs of collective identity are not fully under the control of their creators, one cannot study urban public space only by traditional forms of urban ethnography. As Borer (2006) and others have shown, the meaningfulness of daily life is very much connected to place. Although respondents and informants can convey the nature of affective attachment to particular places, they are not usually in a position to relate the complexity of or multitude of ways that collective identity gets embedded in the built and social environments; much of this communication is implicit and urban dwellers do not always know the full extent of what they know or how they know it. Spatial semiotics using visual data is one effective way of investigating these phenomena. Our analysis demonstrates the value of this perspective for the study of urban life and culture.

NOTES

1. Our archive is available at http://www.brooklynsoc.org.
2. Both expressive and poetic signs imply intentionality. Their difference is primarily in the directness of the intended communication—much like the distinction between literal and metaphoric language. Expressive signs are more direct, and encode both ascribed and voluntary aspects of collective identity (e.g., primary and secondary affiliations). Poetic signs are usually associated with voluntary aspects of identity, because they relate to style, above all, but generally also encode class, ethnicity, and gender indirectly.
3. The additional images described here can be viewed at http://www.brooklynsoc.org/blog/gallery3/index.php/Sociology-of-the-Visual-Sphere.
4. Global commerce helps explain why so many individuals, seeking self-expression of their sense of style, find the same products available for this use.

We are not arguing here that subcultures do not have norms—because obviously they do—but that the difference between primary and secondary affiliations often results in different kinds of signs in urban public space; style, rather than obligation, generates poetic signs.

REFERENCES

Anderson, Elijah. 2011. *The cosmopolitan canopy: race and civility in everyday life*. New York: Norton.

Beauregard, Robert A. and Anne Haila. 2000. "The unavoidable continuities of the city." Pp. 22–36 in *Globalizing cities: a new spatial order?*, edited by P. Marcuse and R. Van Kempen. Oxford: Blackwell.

Borer, Michael I. 2006. "The location of culture: the urban culturalist perspective." *City & Community* 5(2):173–197.

Derderian, Richard L. 2004. *North Africans in contemporary France: becoming visible*. New York: Palgrave Macmillan.

DeSena, Judith N. 2005. *Protecting one's turf*. Revised edition. Lanham, MD: University Press of America.

Fritzsche, Peter. 1996. *Reading Berlin 1900*. Cambridge, MA: Harvard University Press.

Gehl, Jan. 2010. *Cities for people*. Washington, DC: Island Press.

Gottdiener, Mark. 1994. *The new urban sociology*. New York: McGraw-Hill.

Harper, Douglas. 1988. "Visual sociology: expanding sociological vision." *American Sociologist* 19(10): 54–70.

Harvey, David. 1989. *The urban experience*. Baltimore: John Hopkins University Press.

Harvey, David. 2006. *Paris, capital of modernity*. New York: Routledge.

Hawkes, Terence. 2003. *Structuralism and semiotics*. London: Routledge.

Jackson, John B. 1984. *Discovering the vernacular landscape*. New Haven, CT: Yale University Press.

Jacobson, Roman. 1960. "Closing statement: linguistics and poetics." Pp. 350–377 in *Style in language*, edited by T. A. Sebeok. Cambridge, MA: MIT Press.

Jakobson, Roman. 1972. "Verbal communication." Pp. 39–44 in *Communication*, edited by Scientific American. San Francisco, CA: Freeman.

King, Anthony D. 1996. *Re-presenting the city: ethnicity, capital and culture in the twenty-first century metropolis*. London: Macmillan.

Krase, Jerome. 1997. "The spatial semiotics of Little Italies and Italian Americans." Pp. 98–127 in *Industry, technology, labor and the Italian American communities*, edited by M. Aste et al. Staten Island, NY: American Italian Historical Association.

Krase, Jerome. 2002. "Navigating ethnic vernacular landscapes then and now." *Journal of Architecture and Planning Research* 19(4):274–281.

Krase, Jerome. 2003. "Italian American urban landscapes: images of social and cultural capital." *Italian Americana* 22(1):17–44.

Krase, Jerome and Tarry Hum. 2007. "Ethnic crossroads: toward a theory of immigrant global neighborhoods." Pp. 97–119 in *Ethnic landscapes in an urban world*, edited by R. Hutchinson and J. Krase. Amsterdam: Elsevier/JAI Press.

Krase, Jerome and Timothy Shortell. 2007. "Imagining Chinatowns and Little Italies: a visual approach to ethnic spectacles." Paper presented at Le Beau Dans La Ville: Colloque International, Tours, France, November.

Krase, Jerome and Timothy Shortell. 2008. "Visualizing glocalization: changing images of ethnic vernacular neighborhoods in global cities." Paper presented at the First ISA Forum of Sociology, Barcelona, Spain, September.

Krase, Jerome and Timothy Shortell. 2009. "Visualizing glocalization: semiotics of ethnic and class differences in global cities." Paper presented at the annual meeting of the Eastern Sociological Society, Baltimore, MD, March.

Lefebvre, Henri. 1991. *The production of space*. Translated by D. Nicholson-Smith. Malden, MA: Blackwell.

Lofland, Lyn H. 1985. *A world of strangers: order and action in urban public spaces*. Prospect Heights, IL: Waveland Press.

Lofland, Lyn H. 1991. "History, the city and the interactionist: Anselm Strauss, city imagery, and urban sociology." *Symbolic Interaction* 14(2):205–223.

Lofland, Lyn H. 1998. *The public realm*. New York: Aldine.

Lofland, Lyn H. 2003. "Community and urban life." Pp. 937–974 in *Handbook of symbolic interactionism*, edited by L. T. Reynolds and N. J. Herman-Kinnery. Lanham, MD: AltaMira.

Lynch, Kevin. 1960. *The image of the city*. Cambridge, MA: MIT Press.

Osterhammel, Juergen and Niels P. Petersson. 2005. *Globalization: a short history*. Translated by D. Geyer. Princeton, NJ: Princeton University Press.

Pan, Lin. 2009. "Dissecting multilingual Beijing: the space and scale of vernacular globalization." *Visual Communication* 9(1):67–90.

Pauwels, Luc. 2008. "An integrated model for conceptualising visual competence in scientific research and communication." *Visual Studies* 23(2):147–161.

Ritzer, George. 2003. "Rethinking globalization: glocalization/grobalization and something/nothing." *Sociological Theory* 21(3):193–209.

Robertson, Roland. 1997. "Comments on the 'global triad' and 'glocalization.'" In *Globalization and indigenous culture*, edited by Inoue Nobutaka. Retrieved October 2007 (http://www2.kokugakuin.ac.jp/ijcc/wp/global/index.html).

Shortell, Timothy and Jerome Krase. 2009. "Spatial semiotics of difference in urban vernacular neighborhoods." Paper presented at the 9th European Sociological Association Conference, Lisbon, Portugal, September.

Simmel. Georg. 1924[1908]. "Sociology of the senses: visual interaction." Pp. 356–361 in *Introduction to the science of sociology*, edited by Robert E. Park and Ernest W. Burgess. Chicago: University of Chicago Press.

Strauss, Anselm. 1961. *Images of the American city*. New York: The Free Press.

Warren, Carol A. B. and Tracy X. Karner. 2005. *Discovering qualitative methods: field research, interviews and analysis*. Los Angeles: Roxbury Publishing.

Wohl, Richard R. and Anslem Strauss. 1958. "Symbolic representation and the urban milieu." *American Journal of Sociology* 53:523–532.

Zukin, Sharon. 1996. "Space and symbols in an age of decline." Pp. 43–59 in *Representing the city: ethnicity, capital and culture in the twenty-first century metropolis*, edited by A. D. King. New York: New York University Press.

8 Representing Perception
Integrating Photo Elicitation and Mental Maps in the Study of Urban Landscape

Valentina Anzoise and Cristiano Mutti

INTRODUCTION

In this chapter we share our experience in adopting an integrated methodology of two techniques that are normally used separately: photo elicitation interview and mental maps.[1] In the following analysis we will also point out the strengths and weaknesses of said integration as well as the emerging participant–researcher dynamics and their possible development.

Up to now, the methodology known as photo elicitation interview has been used, primarily in social sciences and in particular by visual sociologists, to access people's personal experiences and subjectivity (Wagner 1979; Faccioli 1997; Harper 2002; Hurworth, 2003; Anzoise and Mutti, 2006; Faccioli and Losacco 2008), and mental maps have been extensively tested and used primarily in psychology and behavioral geography. While the American framework is more focused on the relationship between environmental conditions and behavior (Gould 1970; Downs and Stea 1973; Saarinen 1973; Geipel et al. 1980), the European framework is focused on the study of people's orientation, environment perception, and learning, on the representations of space and territories, on the differences between the real and objective world and the subjective one, and on differences between what is represented and what is perceived (Tuan 1975; Bianchi and Perussia 1978).

Nonetheless, despite a common phenomenological perspective in trying to facilitate empathic communication between the subject and the researcher, the integrated use of these two methods has not yet been explored, although they also have common predecessors and background. In particular, the psychological studies developed in the United States in the 1950s influenced all the following studies on perception and behavior, and even the Thematic Apperception Test (TAT), which can be considered the predecessor of the photo elicitation interview, and which was first used by the psychologist Michael Murray (1943), and was extensively adopted by the geographer Thomas F. Saarinen to study group images (1973) and people's perceptions and behavioral reactions to environmental hazards (Saarinen 1966). Furthermore, Helgren's study (1983) to gauge his students' geographic knowledge is one of the first and few published reports on subjects asked to place a location on a map. But all these attempts to "force" methodological and disciplinary boundaries never led to a full integration of the potentialities of both, as we tried to do, instead, in the case-studies presented here.

Vision and Perception

The historian Martin Jay (1988:3) says that "we confront again and again the ubiquity of vision as the master sense of modern era," which consists of three interrelated moments: sensation, selection, and perception. The latter concerning what lingers on people's minds of what grasped their attention and struck their sensitivity.

People's perceptions of environment and space, and the different evaluations they make on different portions of it, generate mental images and mental maps that human beings hold unconsciously, and that provide important keys and cues to understand patterns of representation and structures of people's behaviors and decisions (Gould 1970). The theoretical assumption of cognitive sciences, which do not consider perception as a passive sensing of the retinal image but as an act (Gibson 1979), states that individuals react to their environment on the basis of its perception, and they are able to interpret it on the basis of previous experiences. Moreover, researches in ergonomics and behavioral psychology reveal that individuals perceive the surrounding environment mainly as a visual environment and that, especially in the case of well-educated people, the visual canal collects 80% of our sensorial impressions (Beccali, Gussoni, and Tosi 2003).

Other fundamental insights and reflections about people's differential images of urban landscape have been provided by the urban planner Kevin Lynch (1960:1–2), who stated,

> Nothing is experienced by itself, but always in relation to its surroundings, the sequences of events leading up to it, the memory of past experiences . . . Our perception of the city is not sustained, but rather partial, fragmentary, mixed with other concerns. Nearly every sense is in operation, and the image is the composite of them all. Not only is the city an object which is perceived (and perhaps enjoyed) by millions of people of widely diverse class and character, but it is the product of many builders who are constantly modifying the structure for reasons of their own. While it may be stable in general outlines for some time, it is ever changing in detail.

The assumption that people have a differentiated perception of the environment fostered us to experiment an approach able to elicit, collect, and record information about these perceptions, taking advantage of the uniqueness and potentialities of visual and projective techniques.

Visual and Projective Techniques

Mental maps are the projection on a sheet of paper of the individual's own internal map. The cognitive representation of the environment is a figment of the relationship the individual has with it, which is influenced by different factors, such as personal experiences, the development of cognitive spatial abilities, and the capacity of observation. The image so formed, even if it could remain still in the course of time, is susceptible to small adaptations each time it comes into contact with reality, states Elena Cavallini (2005) according to previous theorizations (Geipel et al. 1980; Gould 1970). Thus, the sketch map reflects the known world, and the map's features, such as the kind of projection, orientation, and size, may be influenced by other maps seen elsewhere, or even by other cultural objects such as paintings, signals, mass media, as well as prejudices and stereotypes, public discourses, or personal experiences. Therefore, from the very beginning it was clear to researchers that we can learn a lot by decoding and comparing mental maps, and the related accounts, provided by groups with different socio-demographic characteristics (Saarinen 1973; Gould 1970; Bianchi and Perussia 1978) because representation and imaginaries of/on spaces and territories are socially constructed and geographically and historically determined (Massey 1995; dell'Agnese 2005).

The introduction of audiovisual and multimedia tools as suitable means for certain areas of research (Wagner 1979; Grady 1996) has to be read jointly with the reflexive turn regarding the construction of knowledge, and the call on social researchers to be reflexive about their own practices of investigation (Bourdieu 1990; Bourdieu and Wacquant 1992; Melucci 1998). Regardless of how much image oriented or equipment oriented (Henny 1986) the visual approach might be, using images constantly requires researchers to self-position themselves, but at the same time it enhances the complexity and inspectionability of the empirical basis of data (Ricolfi 1997).

In this regard, the photo elicitation interview, a semi-structured interview based on images, provides means of verification through which the roles of the researcher and the subject seem sometimes to be reversed. In fact, this technique enables the reflexive researcher to get closer to the object of his/her observation and to constantly challenge his/her hypothesis and assumptions during the interview (Anzoise and Malatesta 2010; Anzoise and Mutti 2006; Faccioli and Losacco 2008; Harper 2002; Hurworth 2003). Furthermore, it also allows the researcher to achieve a co-production of meanings and knowledge. For the above reason, in the case studies that will be presented in the next section, we tried to integrate different research tools and techniques, aiming for the encounter with the subject (object) of our observation, but always questioning the whole research design, in order to avoid the self-deception that may be induced by such powerful tools (Faccioli and Losacco 1998).

CASE STUDIES OF URBAN AND ENVIRONMENTAL PERCEPTION

The description and analysis of the two case studies presented here have a twofold aim: heuristic and interpretative, on the one hand, and methodological, on the other. Besides being conducted in the same urban setting (the city of Milan), what links the case studies are the approach and techniques that have been used to explore two interrelated urban and visual phenomena: territorialization, and perception and imaginary of the city.

Milan, according to the *OECD Review* (2006:11), had "an outstanding past and the radiance of a global city" along the second half of the twentieth century, but now the city is facing the negative consequences of such good economic and commercial performance, for example, low environmental standards, poor quality of life, and loss of city attractiveness. In 2015 Milan will host the World Exposition, and the theme chosen, "Feeding the Planet, Energy for Life," both reveals a growing attention to developing sustainable solutions and constitutes an opportunity to reconnect this hyper-congested urban reality to its countryside. Moreover, it is also an opportunity, in terms of economical advantages, global exposure, and possibility to experiment a cultural and behavioral shift, for the entire Lombardy region, which ranks first for agricultural production and of which Milan is the main city (dell'Agnese and Anzoise, 2011).[2] Hence, in the next few years Milan's metropolitan area will be the stage of great changes that will redesign its material and immaterial infrastructures, its districts, as well as the objectives and uses of its built and green spaces. But how do the different "urban populations" (Martinotti 1993)[3] that flow daily through this city perceive these issues? Do different aggregates of people hold specific images of Milan in their minds?

The first case study has been developed as part of a PhD thesis[4] and is related to the perception and knowledge of the thoroughbred horses training areas and the Gallop Racecourse in Milan, which lie at the end of the city's widest "green wedge." With the expansion of the city, this area became increasingly close to the city center, and it is now tempting the speculative interests of real-estate agencies with a view to Milan Expo in 2015. The second case study has been developed together with the students of didactic labs in the MA course of Sociology at the University of Milano-Bicocca in 2009/2010 and is related to people's perception of the urban green assets in Milan.[5]

First Case Study: The Thoroughbred Horses Training Areas and the Gallop Racecourse in Milano

The first Gallop Racecourse in Milan was inaugurated in 1888,[6] immediately out of the city, in *località* San Siro,[7] and in the following years it strongly influenced the city planning of the northeast area. In addition to fostering the maintenance of an agricultural activity, such as the production of forage for horses, it has constructed the property regime governing the territory (Boatti 2007). In fact, the planning of the training areas in Trenno (1909) and near the Maura farmhouse (1950), and the building of the new hippodrome (1920) limited the erosion of green areas by asphalt and concrete, as indeed was happening all over the city at the beginning of the twentieth century. As a result, Milano's horseracing infrastructures occupy today more than a million square meters. This area lies at the crossing of different neighborhoods (although the whole area is generically called San Siro neighborhood) and includes 700 hectares of fields which are part of the South Milan Agricultural Park. However, a 4 kilometers-long wall, built for security reasons, runs along the private areas of horse training, so in the end, the Gallop Racecourse contributes to keeping these areas hidden from citizens' view. Our first hypothesis was that this wall could be a reason why these areas are less and less known, and therefore less and less present in the collective imaginary of the city. Moreover, both the overshadowing presence of the San Siro football stadium[8] (not only in the same area but also in close proximity to the horseracing areas[9]) and the loss of popularity of horseracing in Italy, could have contributed to this gradual oblivion. In fact, even the image of the whole area, and in some respect also its zoning plans, is more and more shaped by the football stadium, contrary to what was happening until the last decades of the twentieth century, when the San Siro hippodrome was famous all over the world, horseracing in Italy was prestigious and followed by both youngsters and elderly.

The aims of the present research were to start identifying some of the socio-demographic and cultural profiles of the people that know (or do not know) the areas in question, and to develop a consistent and effective approach to access both their environmental perception, what they know and how, and what they think about the territorial planning and the functions of these areas within the city.

Methodologically, the research included only people that, at the moment of the interview, were close to or passing by the horseracing areas (people were stopped randomly by the interviewer, and the interviews were conducted face-to-face on the street). Moreover, we decided to interview people aged fifteen and older (the oldest being eighty-six years old), in order to access both the imaginaries of the very young[10] generations and the memories of elderly people, who may have had the chance to see the areas and the surrounding neighborhoods when their landscape was completely different and their rhythms contrasting with the hectic ones of contemporary Milan. After collecting the socio-demographic data from the interviewees, we asked a few questions about their imaginaries of that area of Milan, the first one being, *When you think about this part of Milan, what is the first thing that comes to your mind?*

Then, people were asked to draw their own map of the area supposedly for a person who doesn't know it in order to get oriented. We also asked them to place all the symbolic places and the ones they prefer and go to most often (labeling the first ones in red and the others in blue, and in both colors if they were both symbolic and preferred). We wanted to see, first, if the horseracing areas were included, and found that in a few cases they were. Moreover, interestingly enough, among the twenty-two interviewees,[11] three did not draw a map, because they either said they were not able to, or because they simply refused. Although in these three cases the mental map is missing, the quote that follows shows that even a refusal can be very rich in information. In this case, the reasons this interviewee shared with us show very clearly that he feels his image of his neighborhood and the meanings he attributes to some of its landmarks (e.g., the stadium) are very different from those of other inhabitants or passersby.

> I wouldn't indicate anything! On the contrary, I would say to everybody to go away from here . . .
>
> *So for you, there is no symbolic place in this area?*
>
> Absolutely not! Nothing! The stadium there, for instance, is not a symbol . . . Can you see there? You won't find anything apart from an enormous imbalance between different social classes. It's absurd!
>
> *So, you won't do the map?*
>
> No, I won't! It would be like inventing a place that doesn't exist. Everybody would say the same, believe me . . . I cannot understand why foreigners come here to take pictures of the stadium . . . What the hell do they see there? . . . Certain people are speculating on it. They earn dough because of the fact that people know San Siro stadium is here. For instance the men who sell the flags of the football teams . . . That is a symbol! They stay there from the morning to the evening to earn money at the expense of . . . yes, that is a symbol! That one! A symbol you pay for, and that it's not for free! That is a symbol . . . the symbol of San Siro.[12]

This quote shows that the identity of a place as well as its definition are always unfixed, contested and multiple (Massey 1994). In this case, although the man acknowledges that the stadium is a symbol for a lot of people and that it attracts a lot of tourists, it seems that for him it cannot really be considered a symbol, and if it is, it is a negative symbol because it is soaked with features and values he considers bad: money, speculation, and so on, which "infected" the whole area as well as the horseracing activities.[13] At the same time, in his opinion, his neighborhood (which is very close to the horseracing areas) is characterized by more relevant issues, such as social inequalities. This issue probably emerged because of the specific place where we held the interview. In fact, at another point of the interview he forecasted violent ethnic conflicts in a few years. Parenthetically, the square where the interview was conducted (which is very close both to the stadium and the horseracing areas) is largely considered the border zone between the rich side of the neighborhood, where also the French school and luxury houses are settled, and the poor one, inhabited mostly by immigrants and marginalized people.

In general, we may say that the methodological approach we adopted allowed the emergence of dissonances and conflicts, although we noticed that the first phase of the interviews—which is the part entailing a more active participation and some easiness with practical tasks (i.e., drawing a map)—is the one that involved more hesitation in the interaction between researcher and participants, especially with elderly people, as the following conversation shows:

> *Can you draw your own map of the area supposedly for a person who doesn't know it in order to help him/her to get oriented?*
>
> Ehm, no . . . what could I say? I can't, I can't draw.
>
> *Do it as you like, draw lines, crosses . . .*
>
> I don't know . . . how could I draw the park?[14]

Drawing represents a unique strategy to explore mind and thought. In fact, it is frequently used to study child behavior (Lowenfeld and Brittain 1984; Malaguzzi 1995), so it is not surprising that people became embarrassed when they were asked to make a drawing, because drawing is different from the mere representation of external space, and it goes beyond it. In their drawings, people sometimes expose things that they would not have revealed voluntarily through other means, or that they are totally not conscious of. It is a synthesis of the intimate and personal world of people and sometimes it is very difficult for researchers to interpret it correctly (Evans 1980), although as stated by Killworth and Bernard (1982:307), "pencil and paper techniques do produce a great deal of inferential data."

Nonetheless sometimes drawings and other visual representations people may provide, for different reasons and constraints, are very remote from the richness in detail of the image of a place people have in their minds. For instance, the following quote, belonging to a sixty-six-year-old lady, who moved to the neighborhood a year and a half before the interview, reveals a strong affective bond to this new place of residence, emotionally connected it to her childhood.

> *When you think about this part of Milan, what comes to your mind?*
>
> It comes to my mind that I like it a lot, because when people ask me where I live the first thing that I say is that I live in the countryside! I

spontaneously say that I live in the countryside, because I feel so good staying here in the green and in the quietness . . . this area is really calm and inhabited by very simple and respectable people . . . this area calls the countryside to my mind, so if anybody should ask me for an advice I would suggest everybody to come live here. I really love to live here, in the countryside, it makes me go back to when I was a child, because I spent my childhood in the countryside with my parents and my brothers."

So this area of Milan reminds you of the countryside, while other areas don't?

Yes, others do not . . . for instance the neighborhood where I used to live was very nice and full of green, but there were also many houses, buildings, concrete, etc. . . . Here, with all this green, I feel like I live in the countryside.[15]

Nonetheless, when the lady was asked to draw the map of the area, she felt a little uncomfortable. She used the whole sheet of paper, and divided it to four parts, of which her neighborhood occupies two. On the map she wrote down only the name of her neighborhood and the names of two nearby parks, but she was not able to translate to visual representations most of her verbal descriptions of the neighborhood. However, her map still shows her deep perception of the neighborhood as characterized mainly by green areas, which she valued so much.

The next quote, which belongs to an eighty-year-old man, who was born in the neighborhood and has always lived there, shows other social and individual processes related to the emotional connections between physical environment and human beings, which Y Fu-Tuan (1974) named "topophilia."

What kind of drawing do you want me to do?

Something like a map, where you indicate the places you go to in this neighborhood . . .

No, no . . . I'm barely able to draw my signature.

But we only ask you to draw a sketch map of the neighborhood . . .

The neighborhood, the neighborhood . . . what are you talking about? Don't talk to me about these things . . . What are you on about? . . . In my time you had to go to work in the fields to eat. Oh Jesus, I started to go to school when I was seven, not six, because when I was six years old I was sick, then I had to repeat the year, and then, then . . . in the morning I used to go to school and in the afternoon I had to go to work in the fields if I wanted to eat . . .

Did you use to live here?

Yes, I was living in the Cottica [a farmhouse in the neighborhood]

I want to show you some pictures now . . . have a look at them and please tell me what you see there, if you recognize anything, if you know where these places are . . .

Eh, [he speaks in Milanese dialect] these are the racecourse stables, at the end of Trenno. I used to work there; I worked there for thirty-five years.

Really? And what were you doing there?

I did the maintenance of the racecourse.

The one in the hippodrome?

Yes, then they decided to cut the staff, I worked there thirty-five years and they fired me . . . [16]

    This interview shows that some past images and experiences can have such a strong effect that people are hardly able to refer to current conditions and features of a place. Moreover, the culture of a society, its educational practices, models, and principles, are all crucial factors in the recognition and acceptance of ourselves, both in the surrounding physical space and in one's own feelings and thoughts. It means that, for instance, when a territory that was once familiar has changed so much because of urban expansion, being asked to draw it (e.g., the neighborhood) facilitates the emergence of individual perception and dissonances referring to the social as well as the territorial processes which were involved in it (e.g., conflict, and urban and labor change).
    Lynch states that people's need to recognize and pattern their surroundings is crucial, especially in the process of way finding, and it also has high practical and emotional importance for the individual. Then, he points out that the environmental image is "the product both of immediate sensation and of the memory of past experience, and it is used to interpret information and to guide action" (Lynch 1961:4). Thus, occasionally, when people do not find correspondences between their past and current environmental images they may feel legitimized to declare a sort of incapacity or an inadequacy to express themselves in relation to the surrounding and current environment. However, when conducting social research we must be aware of our own interventions with our research tools and of the ways these give rise to or uncover such reactions. After all, being asked to draw a map certainly is not a common thing to do in the everyday life of regular people.

Analyzing the Mental Maps

We analyzed the nineteen maps that have been compiled considering the following features:

- *Scale*: the area included in the drawn map. Only three maps represent a fairly extended portion of the city, while the other sixteen focus only on a restricted area. This is quite normal because they were asked to draw a map of the district they were in at the moment of the interview.[17]
- *Perspective*: aerial or frontal view. Sixteen maps had aerial view, and three maps had frontal view.
- *Orientation:* cardinal directions indicated or identified. People were not asked to put the north on the map, and among nineteen only two of them did it spontaneously writing it on the map. Sanders and Porter (1974) state that when people are drawing they identify the upper part of the sheet of paper with the north, and this was the case also for most of these maps, but we wanted to see if interviewees' orientation could be influenced by their position at the moment of the interview.
- *Styles of Representation:* graphic solutions adopted to connect, identify and delimitate spatial elements. We have ten maps drawn in what Bianchi and Perussia (1978) define a "zone style," which means that they give a detailed, almost topographic, representation of streets and squares, five "radial maps" (i.e., lines that connect different points), two "round maps" (i.e., an area within a closed circle with some points more or less connected with each other) but we also have one "minimal map," which we define as a map where points are indicated with names or circles which are not connected at all.
- *Typologies of Representation*: graphic elements used to draw points and landmarks, that is, texts/names, symbols, lines and circles, and so on. In all the maps there are texts with lines, squares and circles, but four of them also include some symbolic elements, such as houses seen from the front to identify the interviewee's home or residential areas.

The integrated analysis of all these features could provide a good measure of the complexity and richness of the map. While they were drawing the map, sometimes the interviewees themselves made explicit how they use, and process spatial information and landmarks to get oriented.

If you had to put the north on the map, where would you put it?

The north? You know I have no idea!? [laughing] I have some problems with the cardinal points . . . Uhm, west part . . . because San Siro is in the west part of the city, at least I know this . . . so north would be . . . here? . . . I don't know.

Representing Perception 139

So, what makes you say the stadium is here? What did you take as reference point?

Well, as reference point I put myself in perspective with the stadium, as if I gave it my shoulders, so here there is the big avenue . . .

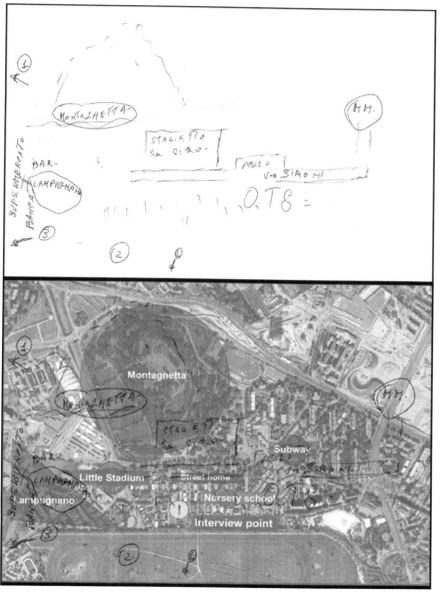

Figure 8.1 Mental map overlapping with the geographical map extracted from GoogleEarth.

So you are projecting yourself in the space referring to the stadium position?

Yes, exactly. As if I were looking at the sheet of paper and the stadium is on my back.[18]

This quote shows that people do not necessarily orientate themselves through identification of the north, which exists as a convention, but they often use their body (turning or projecting themselves mentally) to match their conceptualization of space.

As far as we can understand from the drawings and the conversations carried out during the drawing of the maps, ten maps were oriented on the basis of certain points of references that were present and visible on the ground, seven others by mentally projecting oneself in the space, sometimes using elements not visually accessible (i.e., their home), and two others—in maps that covered quite a wide portion of the city—adopted a cartographic logic to represent their general geographic knowledge of the territory.

In the next step of our research we overlapped the mental map with a geographical map extracted from GoogleEarth (Figure 8.1). We developed a Flash application to check, with a transparence command, if there were any correspondences, or not (Figure 8.2). The aim of using the GoogleEarth Map was to have a conventional way to represent space as term of comparison. To do so, we took into account and then fixed only one of the main variables: the positioning of landmarks on the mental map, while scale and orientation of the GoogleMap were adapted by us so that they

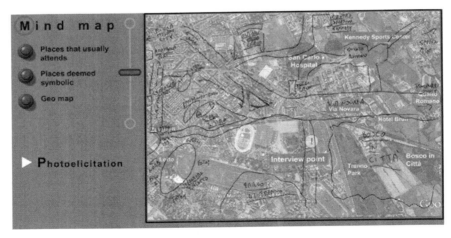

Figure 8.2 The interface of the Flash application that generates hybridization between the mental map and the geographical map showing the correspondences between the elements present in the two maps.

would best fit those of the mental maps. For this adjustment we were mainly assisted by the landmarks placed on the mental maps, and by their explanations where interviewees made explicit they were mentally projecting themselves on the surrounding space to get oriented, or taking some points of reference that sometimes were not visible in the location but present in their mind (e.g., their home).

Consequently, we tried to establish an index of correspondence by counting the places indicated in the drawn map and those that were correctly corresponding.[19] The outcomes show a high correspondence in twelve cases, medium in seven and low in none.

Analyzing the Photo Elicitation Interviews

The photo elicitation interview, which followed the drawing of the mental map, was constituted by five images, and started this way: *I want to show you some pictures. Do you recognize this place? What comes to your mind looking at it? Are you able to place it on the map you have just drawn?*

The set of photos shown (Figure 8.3) was chosen both to see if people were able to recognize different landmarks related to the horseracing activity[20] and to locate them on the map, as well as to stimulate a deeper discussion on what the presence of these elements within the city means to them. Subsequently, and particularly in the case of Photo 1 which was shot inside the training area where the old stables are, portraying a private place not visible to everybody, the aim was to see if people could recognize (or at least attribute) it as part of Milan's landscape.[21]

For the interviewees, the photo elicitation interviews provided interesting elements to reflect on, and it triggered their memories and imagination. In fact, after this phase of the interview some of them returned to their drawn maps to add further landmarks. This occurred especially after commenting on Photos 4 and 5, that some interviewees recognized only afterwards as unique elements of the neighborhood's landscape.

Although those who are not residents of any of the neighborhoods of the area had difficulties in correctly placing the photographs (especially Photos 1 and 4), all interviewees' comments and reactions to the pictures somehow casted light on the variables and environmental features which could be more influent on people's visual selection and perception. On the one hand, most of the people who could recognize the majority of the places are residents of the districts which border the racing areas and felt free to express openly their opinions about the changes that are occurring there, or about the lack (or loss) of a clear idea of what is their daily use. On the other hand, even people who did not recognize all places in the images, but who are somehow and for some reasons hanging out around the racing areas, disclosed their imaginary of them and of Milan in general, as the following quotes, from three different interviews, express very well:

142 *Valentina Anzoise and Cristiano Mutti*

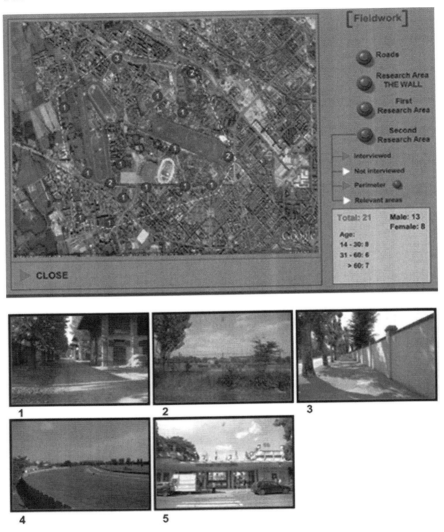

Figure 8.3 Top: research area and interviews' localization;[22] bottom: photo elicitation set.

I don't know where this is . . .

But in your opinion, could it be here, in Milan?

Yes . . . it could be, but I don't think so . . . No, I don't know, but no, this is definitely not Milan . . . [23]

***

Representing Perception 143

I follow my intuition . . . these should be the old houses of the hippodrome, but I don't know on what side they are, or if they are really there, because they look more decadent than I imagined . . . because it's so many years that I don't go there anymore, but they could be the old stables of the hippodrome . . .

Would you locate them somewhere on the map?

Well . . . maybe near Zavattari square, where the "Sole24ore" headquarter is . . . in San Siro neighborhood.

And do these places bring up any particular thoughts?

Yes, that Milan could be beautiful . . . It has to be regenerated . . . because, well maybe it's not Milan but Turin, I have no idea, but the area of the old houses and stables of the hippodrome, that is exactly towards San Siro, is decaying I think, and it would be worth, indeed, to regenerate.[24]

***

Looking at this picture what comes to your mind?

A wall.

Do you know what's behind that wall?

No

Have you ever walked along this street?

Yes, by foot, but on the other side of the street.

Have you ever noticed this wall?

Yes, and I just thought "a wall," as that one there . . . [we are in Pinerolo street and the interviewee indicates another part of the wall surrounding the hippodrome]

And . . . what's behind that wall?

The Nothing. The end of the world [ironic] . . . No, I don't know, I have never wondered about that . . .[25]

These quotes, and the last one in particular, clearly show there is no imaginary at all about the horseracing areas, although big and bordering another

144 *Valentina Anzoise and Cristiano Mutti*

landmark relevant to people (i.e., in the last case, the stadium, because this interviewee is a football supporter who was going, and often goes, to watch matches at the San Siro stadium).

In the following tables we tried to summarize some provisionary outcomes from the analysis of the photos' recognition. In the rows we listed the photos used in the photo elicitation interviews and in the columns we counted the cases in which people (a) were able to recognize quite precisely the spatio-temporal elements represented (where and when the picture was taken) or (b) had any knowledge about the place. The last two columns of Table 8.1 indicate whether the interviewees were completely unable to recognize the object of the photo or were convinced that the picture had been taken somewhere out of Milan. Analyzing these data, we can say the landmarks portrayed are, for the majority of the interviewees, recognized as elements of the area, although the comments people made on them during the interview do not reveal an equally rich knowledge and interest in their function and use. Table 8.2 simply indicates how many people could (or could not) position the photos correctly on the mental map they had drawn or refused (or gave up) locating it[26]. Considering that only nineteen interviewees drew the mental map, and that only Photos 3 and 5 were correctly located by more than half of them, what emerges is that apart from the photos taken "looking at the outside" of the horseracing areas (i.e., the wall and the entrance of the Gallop Racecourse), the other photos

Table 8.1 Photos' Recognition

Photo	Recognition of spatio-temporal features	Recognition of it just as representing an element of the area	No recognition	Thinking the place is not located in Milan
n. 1	8	11	2	1
n. 2	13	1	8	0
n. 3	17	4	1	0
n. 4	6	11	5	0
n. 5	22			0

Table 8.2 Photos' Positioning on the Mental Map[27]

Photo	Positioned in the correct place	Positioned but in a wrong place	Not positioned at all
n. 1	6	7	6
n. 2	8	5	6
n. 3	11	2	6
n. 4	3	5	11
n. 5	14	0	5

which were taken "looking inside" the areas make it difficult for people to orientate themselves, creating a sort of fracture in people's mental map.

A last consideration regarding the predominant image attributed to that part of the city (i.e., the first thing that comes to their mind thinking about the area): only four interviewees[28] indicated the hippodrome and the neighborhood's social inequalities,[29] five the San Siro stadium, and twelve people the green assets. These outcomes, apart from highlighting that the horseracing areas are less and less known and present in the collective imaginary of the different urban populations (Martinotti 1993) of Milan, show that for more than half of the interviewees the main and most valued characteristic of the area is the presence of green. This last outcome was a further inspiration to test this methodology in the following case study on the perception of green areas in Milan, considering the city scale.

Second Case Study: Urban Forests and Green Areas in Milan

Recent studies on human perceptions and behavior concerning urban forests and green areas have shown the complexity and the multidimensional character of the culture–nature relationship in the city and the important functions they perform (Chiesura 2004; Sanesi et al. 2006). In fact, beside environmental, aesthetic, psychological and health benefits, they can provide other benefits: they encourage the use of outdoor spaces and increase social integration and interaction among neighbors (Coley et al. 1997).

Consequently, in this second research we focused on the exploration of the perception and the "public images" of green areas held by Milan's inhabitants and city users. We started with the assumption that stress is a common aspect of urban daily life and that Milan is broadly considered the most stressful as well as grey city of Italy. Many researches show that compared with urban scenes with no vegetation, natural environments with vegetation and water induce relaxed and less stressful states in their observers (Chiesura 2004; Schroeder 1991). Therefore, we wanted to find out whether people who live in Milan are familiar with the city's natural and green areas, and if they function as "natural tranquillizers" for Milan's populations, or whether they provide other functions, if any.

The research was first developed together with twenty students from the MA program in sociology attending the didactic lab on visual techniques for social research, held in 2009–2010.[30] They were asked to reflect on Milan's green areas, and then, as a practical exercise, to collect visual data on the following dimensions of green assets (which also correspond to the main uses and functions of green areas within urban environments):[31]

- Green areas for sport and leisure (C1)
- Exemplary green areas (e.g., botanical garden, didactic farm and schools' vegetable plots) (C2)

- Peri-urban agriculture (vegetable gardening in urban environment) (C3)
- Decorative/ornamental green (including the green used for street decorations) (C4)
- Private/hidden green (e.g., private gardens, terraces) (C5)

After collecting the visual information, we selected together with the students a few pictures for each dimension[32] and then set the structure of the second phase of the research: the photo elicitation interviews. The process of choosing the photos was guided by their degree of iconicity (Mattioli 1986) but also by their efficacy (how much they invite people to talk) and effectiveness (how much they invite people to talk about that specific topic) (Faccioli and Losacco 2008), which were also tested in a few pilot interviews. In the end, because the purpose of the exercise was didactic and the main aim of the research was to test a methodology, we decided to conduct the interviews with two different set of photos,[33] and each group of students[34] had to test both. We collected fifty-seven interviews,[35] carried out in five different areas of the city. Among the interviewees, twenty-four were natives of Milan, thirty-three were living in Milan, and twenty were simply passersby in the neighborhood where the interview took place.

As in the previous case study, the collection of socio-demographic information[36] was followed by a few questions aimed to explore people's perception and imaginary of Milan (i.e., its symbols, the main green areas, etc.). We then asked research participants to draw a map and include the following elements in it: (a) what they consider as the symbols of the city and (b) the city's green areas, labeling with a different color (in red) the ones they use to go to, and trying to connect all the elements to each other as if it was a map for orientation.

Among fifty-seven interviewees, eight refused to draw the map, and only four people made a drawing with frontal perspective, one of which was an allegorical drawing (Bianchi and Perussia 1978).[37] As in the previous study, most of the people drew the map from above, then seven used a large scale of representation (i.e., the whole city), nineteen used a medium-size scale (i.e., a portion of the city but quite wide), and nineteen used a small scale (e.g., a district). The style most frequently used was the "minimalist map" (seventeen maps), then "zone style" (twelve maps), "radial map" (nine maps), and "round map" (seven maps). The high number of "minimalist maps" could be the result of the few minutes people were willing to dedicate to the interview conducted in the street. The scale of representation seems to have influenced the map orientation. The majority (forty-one people) indicated north on the top, in many cases adopting a perspective from above and a relatively wide scale of representation.

Fourteen interviewees used a mental projection of themselves in space to draw the surrounding elements and areas. This happened both with people representing a limited portion of the city or a much wider one, as it is well

Representing Perception 147

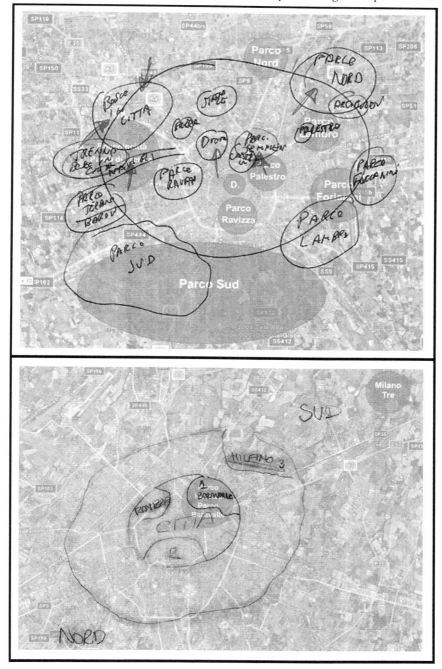

Figure 8.4 Top: mental map with high correspondence. Bottom: mental map with north and south completely reversed made by interviewee GR3-Int05-B.[38]

shown in the map drawn by a teenager[39] interviewed in the city center (Figure 8.4, bottom). The interview with her reveals that her map projection was made physically looking toward her home, which is located in a village in the south of Milan (indicated with "Sud" on the map).

Only two people clearly used visible elements present in the surrounding environment to orientate themselves and draw the map (and both used a small scale), while thirty interviewees used their previous and accumulated geographical knowledge of the city, and thirteen declared they were mentally projecting themselves in the space in order to be able to draw the map. We noticed a high correspondence[40] between the GoogleEarth Map and the mental map in twenty-four cases, medium correspondence in nineteen maps and low correspondence in two maps (however, each map was different in the number of points marked; see Graph 8.1).

The predominant image of Milan that emerged in the interviews is that of a chaotic place, full of job opportunities, businesses, and shopping, but not very livable and green. It is interesting to note that when we showed them the sets of photographs seven people recognized the Sempione Park, located at the center of the city and very close to the Milan Cathedral and the Sforza Castle (considered the main symbols of Milan by most of the interviewees), as another symbol of the city. Nonetheless, compared to other European and North American cities, Milan is perceived as grey, and the existing parks are considered just little spots. In fact, when asked to indicate the green areas they know and go to, people mainly indicated two typologies of parks: the very central and then the peripheral and peri-urban

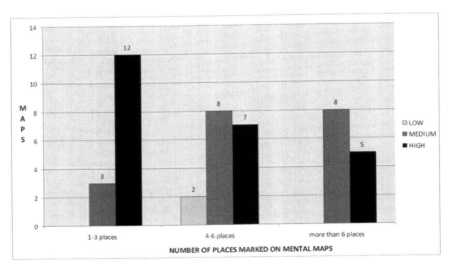

Graph 8.1 Degree of correspondence and number of places marked on the mental maps.

ones that are the only fairly large ones. It seems that Milan, in the eyes of many of the interviewees, is made of cement and asphalt, as the comment of a student of the Polytechnic of Milan, daily commuting to Milan, points out very well:

> *If I asked you to indicate the green areas of the city, which ones would they be?*
>
> Absolutely none, there's no green in Milan, as far as I know . . . [41]

Certain people were positively surprised to see pictures of parks, gardens, vegetable plots within the city, because the photos made them aware that their image of Milan with no green was not completely accurate.

Another important variable which emerged as crucial when drawing a map was people's mobility patterns. For instance, commuters and "city users" (Martinotti 1993) found it quite difficult to draw the map of the city, although they make use of it and its services almost every day (especially in the case of commuters). The mental maps collected show that ten people barely knew the city, and drew only a map of the center or just a little portion of Milan (e.g., the district they go to study or work), but in some cases they indicated landmarks referring to their patterns of mobility (i.e., railway station, underground, etc.). Nonetheless, even for Milan's inhabitants, means of transportation are crucial elements for the city perception and knowledge, as the map drawn by a taxi-driver[42] (Figure 8.4, top) and the following comment of an immigrant from Bangladesh, illustrate very well:

> These are bus lines 73, 45 and 27 . . .
>
> *And where are the parks?*
>
> The parks are here, at the edges, here and there.
>
> *Can you draw them?*
>
> These are bus lines 66 and 62, and here is bus 45's stop, here buses 45, 66 and 62 go straight . . .
>
> *And they go where the parks are?*
>
> Here there is a park.
>
> *Ah ok, this is Cadore Park?*
>
> Yes.

And this one is Solari Park?

I don't know very well but here there is a playground for children, here is the exit, the street, and here is where bus 92 passes by. Here there is another exit, and we have buses 29 and 30, this is the park, and over there you have tram 14 . . . Here you also have benches and there is a swimming pool near there . . .

Do you go to these places?

No, not always.[43]

Figure 8.5 The two sets of photos (A and B) used for the photo elicitation.

Representing Perception 151

The last questions focused on the set of photos (Figure 8.5), showed one by one. People were asked to comment on them and then to place each photo on the map they had drawn.

The following excerpt (as others) from the photo elicitation interviews reveals people's ideas about what they considered to be green areas, what these areas should look like, and what the local government of the city should take more into consideration (people's needs, maintenance, the history of the city, city livability, etc.):

> I don't know where this place is . . . it seems new, it's not the classical square . . . this square seems to me . . . well, not inhuman, but I think it doesn't prompt people to go there.

Figure 8.5 The two sets of photos (A and B) used for the photo elicitation.

Can you indicate the elements that do not prompt you to go there?

> Well, for instance this hidden area . . . I can't see benches where to sit . . . it's just a square to pass by, not to relax. I think that this is the problem of Milan. I mean, this city is perceived as a city just to pass by, to work, to take the tram or the train to go home. It's not a city thought of as for living, for people. If I go to this square . . . where can I sit? Maybe under this tree? And what about children? I can't see anything for playing there . . . so, I do not consider this area a green area, it's a square, or just a part of a street, but not a green space.[44]

The last pictures, which represent the horseracing areas (Photos 6A and 6B), were used in both sets of photos in order to connect the two researches that were exploring interrelated issues. Both photos worked very well. In fact, although only a few people (thirteen) attributed them to Milan's landscape, and even fewer people (three) were able to recognize and locate them on their map, and some people even thought they could be portraying Milan's rural past,[45] these two photos served as a very effective stimulus. Interviewees commented on both photos at length, sharing both their personal feelings and their ideal image of the functions that green and nature have in the city. Interestingly, it is exactly the picture that people rarely recognize as representing a place in Milan the one that stimulates people's more dominant image of Milan (modern, grey, stressful, etc.) and that encourages them to talk more freely and emotionally. The following excerpt illustrates this quite well:

> This must be part of a park . . . maybe the north park . . . because there's an area with stables there . . . I don't know . . . but anyway, I like it.
>
> *Why do you like it?*
>
> Because it gives me a feeling of calmness, it gives the idea of an inhabited place, calm but not isolated, because there are signs of human activity, you can also see aerials . . . I think that people have lived there recently.
>
> Then I also like the light of the sun and these open spaces and paths where you can walk that are covered by leaves. It seems a human-sized place to me . . . and then I don't know, I like the architecture, relaxing but very colored, while in the other pictures the grey prevailed . . .[46]

CONCLUSIONS

The main goal of these researches was to examine if and how an integrated and participatory methodology on a relevant issue such as urban and environmental perception could work. Following this examination, one of our findings deals with the procedure of selecting the "right" photos to be used in the photo elicitation interviews. At the beginning of the process we had long debates over this issue, in class and in an online forum we specifically set for the didactic lab. We then asked the students to comment, in writing and orally in class, on the photos taken by others, ranking them according to their degree of iconicity, in this case the capacity to visually represent a certain category, evaluated in terms of efficacy and effectiveness.[47] At the end of the process, because there were contrasting opinions on the efficacy of certain photos, we jointly decided to use two different sets of photos in order to compare their efficacy and the efficiency. We found this process to be useful for it allowed us to pre-examine images already within the group (thus eliminating photographs which the majority of participants defined as problematic for various reasons), and it also allowed us to examine our final choices during the research itself (by using the two sets of images) and implement the results in a follow-up research.

In this follow-up research we further developed the exhaustiveness of the visual indicators in a third case study (the outcomes and the final analysis of which are not yet available), using a composition of photos that could reflect the complexity of the different dimensions of the general issue. We also decided to simplify the set of questions, in order for people to concentrate only on the map and the photos, and then to isolate, as much as possible, the influence that the location of the interview can exert.[48]

Another outcome of this research emerged from the double request made to people to both draw their mental map and to say if they could recognize certain photographs and place them on their map. This task allowed the exploration of other dimensions regarding the kind of geographical and visual knowledge people have of the territory and how they gather such knowledge. Moreover, having them drawn something with their own hands facilitated the access to their subjectivity: their needs and the problems they perceive and the connections they see between the different elements (landmarks) and the functions they have; for example, in the second case study we learned that regardless of whether they know Milan's green spaces, they are not satisfied with them. It seems they complain about the quality of city life in general, so they express the desire for well maintained and fully usable areas, but they also attribute a strong value to the restorative function, and the sense of peacefulness and tranquility that green and nature can provide even if not usable, or usable only by certain groups (as in the case of private gardens, vegetable plots, green used for street decoration, etc.).

Furthermore, we found that different urban populations may have developed differential perceptions of the whole city and of its districts. For example, people who use different mobility patterns and make different "uses" of the city[49] seem to develop different kind of knowledge and representations of the city.

It is important to note that the aim of both researches shared here was not to evaluate people's mental map in terms of their mental cognition (we were not interested in the degree to which people translate their mental understanding of space to a graphic representation). We did, however, asked to compare these maps (assuming they provide a partial representation of the way people perceive their environment) to geographical maps (which provide a conventional way to represent space). In fact, one person's cognitive understanding and representation of his/her surrounding environment results from many direct and indirect experiences (also including many other maps).

Trying to find out the level of correspondences between the mental maps and the geographical maps was not an end in itself. The problem of comparability has already been raised at the epistemological level by other scholars who state that these maps have some analogies in their use but not in their intrinsic nature, and do not have the same geometric features (Blaut, Mc Cleary and Blaut 1970). By taking into account only one of the main variables —the positioning of the landmarks—we nevertheless wished to emphasize this analogy. The other variables (scale and orientation) were adapted by us. This approach highlights the fact that people, when drawing a map, use different ways to orientate themselves. This can be seen in the case of a woman who drew her map as if everything was to her back, but when she indicated her son's school she drew it as if it was in front of her (was she influenced by a daily habit or perspective?). What happened is that in her mental map the school is exactly opposite to where it "really" is (i.e., on the geographic map). The school is reversed like in a mirror whereas the other places are corresponding. This is not the only case in which the researcher had to interpret and try to fill some gaps and decode some inconsistencies. This situation is very interesting because it forces the researcher to question how the interviewees interpret their own lived space, and all the visual information surrounding them.

To conclude, there are still some open questions, such as the possibility of gaining access to interviewees' subjectivity and their ways of seeing the world using visual representation, or the possibility of inferring from mental maps to how people build their images of their environment, as well as the relationship between what people imagine (stereotypes) and what people see (cognition) and eventually draw. However, we think it is worthwhile to continue testing integrated approaches to study complex phenomena and processes (such as getting oriented, the becoming of habits, etc.), which fall in the realm of cross-disciplinary research.

ACKNOWLEDGMENTS

We wish to thank our students for their enthusiasm in carrying out the researches with us; Stefano Malatesta, a brilliant geographer, for his critical observations on the methodology and on the following analysis; and Camilla Gamba and Alfonso Vinassa de Regny for the English review of the article.

NOTES

1. We first presented this approach, and the explorative researches conducted to test its effectiveness, in a paper presented at the XVII ISA World Congress of Sociology, Gothenburg (Sweden), 11–17th July 2010.
2. For more detailed data on Lombardy agricultural production: http://www.agricoltura.regione.lombardia.it.
3. For the urban sociologist Guido Martinotti (1993), the factors that contributed to the evolution of the metropolis let also emerge different "urban populations", i.e. aggregates of individuals that temporarily share everyday practices and create peculiar space-time patterns (i.e. inhabitants, commuters, city-users, metropolitan businessmen), which led to a further differentiation in the city functions and in the uses people make of it. For Martinotti, the increased mobility of people, combined with greater income and leisure, allowed the differentiation of a third population: the "city users", which is composed of people going to central cities and using private and public services (from shopping, to restaurants, public areas, etc.). The size of this population is growing and very difficult to assess, although its presence is transforming the city structure
4. Written by Cristiano Mutti.
5. The project was coordinated by Valentina Anzoise and Cristiano Mutti, and continued, with some methodological variations, in 2010/11.
6. Milano's second Gallop racecourse, where this research was conducted, was inaugurated in 1920.
7. At the time it was not a district, as it is today, but one of the many little villages in the outskirts of Milan.
8. The Stadium was expanded in occasion of the FIFA World Cup in Italy in 1990.
9. From now on we will refer to the Thoroughbred horses training areas and the Gallop racecourse simply by "racing areas", since it's not a neighborhood but a delimited area lying at the crossing of different neighborhoods.
10. In Italy, the ethical code for conducting interviews with children and young people requires parents' permissions only for children under 15 years old or for young people under 18 years old if the interviews refer to sensitive issues, which was not the case in the present research.
11. Thirteen men and nine women, of which, eight were 15–30 years old, eight were 31–60 years old and six were above 60 years old. Only eleven were born in Milan and sixteen lived in the neighborhood at the time of the interview.
12. Interview 7 (June 2009), sixty-six years old man, living in San Siro neighborhood, which borders the racing areas, since thirty years.
13. See his comment on Photo number 5, below.
14. Interview 5 (June, 2009), seventy-seven years old lady, born in Milan and living more than four decades in Gallaratese neighborhood, which is very close to the racing areas.
15. Interview 6 (June 2009), sixty-six years old lady, born nearby Milan.
16. Interview 16 (March 2010), eighty year old man, born and still living in Lampugnano neighborhood, which borders the racing areas.
17. In the second case study described in this paper we investigate more the role played in representations of space by the scale people decided to use in drawing the maps (Ewoldsen *et al.* 1998; Sholl 1987).

18. Interview 21 (June, 2011), 19 years old boy, who was born and still lives in San Siro neighbourhood, which borders the racing areas.
19. We calculated a simple ratio between the number of places indicated on the mental map and the correct matches using the following classification: High—up to 65% successful matches; Medium—between 25–65%; Low—under 25%. These values should be weighted by another relevant value for the analysis: the overall number of places that were indicated on the maps (High: up to six places; Medium: between four and six places; Low: three or less).
20. Photo number 1 shows the training areas from inside, where the old stables are; Photo number 2 shows Trenno Park, located nearby the training areas which are, in fact, visible from certain points of view; Photo number 3 shows the wall surrounding the whole area; Photo number 4 shows what is beyond the wall (in this case the Maura training course); Photo number 5 shows the entrance to the Gallop racecourse.
21. The set of photos was pre-tested and in fact the order and certain images were changed following that test. For example, Photo Number 3 replaces a photo which showed a graffiti on the wall We replaced in with a photo portraying a blank piece of the wall, because people were commenting more on the graffiti than the wall, and placed before Photo Number 4 to see if people were able to connect what is in front and what is behind the wall.
22. The numbers on the map indicate how many people have been interviewed in that point of the area.
23. Interview 1 (June 2009), conducted with a nineteen years old boy recently moved to live in San Siro neighborhood, which is close to the horseracing areas, commenting Photo number 1.
24. Interview 14 (October 2009), conducted with a forty-six years old lady, born in Milan and living for thirty years in Trenno neighborhood, which borders the horseracing areas, commenting Photo number 1.
25. Interview 10 (July 2009), conducted with twenty-one years old man living in a town nearby Milan, commenting Photo number 3.
26. Had we been working on a statistical sample, we could have presented further outcomes by the cross-tabulation of this information with the socio-demographic characteristics of the interviewees or the specificities of the photos themselves. In this paper our main goal is to elaborate on the methodology we employed and show its potentials and limitations.
27. Compared to the geographical map.
28. Interviews 4, 8, 11, 19 (made between March and July 2009). Three interviewees were males and one a female, and among them two were born in Milan, the other two were not; two are living in the neighborhoods nearby the horseracing areas and the other two are not, although they have some familiarity with the area because one has family in a nearby village and the other one is a supporter who goes quite often to the San Siro Stadium.
29. The district is actually characterized by strong socio-economic divisions and inequalities, on one hand it is inhabited by low-middle class citizens and immigrants, and on the other hand by very upper class citizens (football players, professionals, entrepreneurs and businessmen, etc.).
30. Valentina Anzoise and Cristiano Mutti held the didactic lab at the University of Milano-Bicocca (Italy) between 2004 to 2011.
31. In the following pages we will use C1, C2, C3, C4, C5 to refer to the five categories we used for the urban green areas we studied (classified according to their use and functions).
32. The process was quite difficult and long since we had hundreds of photos to choose from. We tried to make it easier by using a website onto which we uploaded all the pictures, shared them and commented on them.

33. Each set of photos was composed of five photos representing the five categories plus one photo taken from the first case study. This way we tried to connect the two studies, not only methodologically but also conceptually, since they were exploring interrelated issues
34. There were five groups working on this project, each composed of no more than four students.
35. Of the fifty-seven interviewees, twenty-four were interviewed using set A, and thirty-three using set B. Twenty-eight of all interviewees were men and twenty-nine women. Twenty-five people were 15–30 years old, twenty were 31–60 years old, and twelve were more than 60 years old.
36. Age, gender, education, profession, place of birth, neighborhood of residence, reason why he/she was in that part of Milan at that moment (if not resident in the area).
37. In the case of allegorical maps people express a concept through something concrete. In this case people asked to draw something (e.g. a city or part of it) decided to draw a specific, and very concrete element, which expresses their idea of what they have been asked.
38. This code means that this interview is the fifth one, made by Group 3 (GR3) using the set of photos B.
39. GR3-Int05-B (January 2010), sixteen years old girl, living in Rozzano a village bordering the South of Milan.
40. About maps' correspondence see note 19 above.
41. GR1_Int07_B (January 2010), twenty-five years old male, living in a town near Varese (around 75 km from Milan).
42. The map he drew shows he knows very well the city, and in fact in his map both the correspondence and the number of places marked are high.
43. GR3_Int2_B (December 2009), twenty-six years old male, born in Bangladesh and now living and working in Milan.
44. GR2_Int02_A (January 2009), twenty-seven years old female, born in Barcelona (Spain), living in Milan since almost one year to do the fieldwork for her PhD thesis, commenting Photo No.4B.
45. When the city was full of farmhouses. Actually, nowadays, this side of the past of Milan is almost unknown, especially by the youngsters and by those that are not native of Milan.
46. GR1_Int01_A (December 2009), twenty-four years old male, born in city very far from Milan (Bari, in Puglia region) but now studying in Milan, commenting Photo No.6A.
47. Faccioli and Losacco (2008) first described these two features which are related to the degree of iconicity or the reliability of a photo used as a visual indicator (Mattioli 1986).
48. In the two case studies detailed here people were interviewed on the street, while in the third case study we interviewed people at home or in another indoor place after having set an appointment with them.
49. For example residents, commuters, passers-by, tourists, etc. make certain and different uses of the city, which also differ in time, because they can be daily or systematic or sporadic, etc.

REFERENCES

Anzoise, Valentina and Cristiano Mutti. 2006. "Guerra e trasformazioni socio-territoriali. Una ricerca audiovisuale sulla città di Mostar." Pp. 36–51 in *Violenza senza legge. Genocidi e crimini di guerra nell'età globale*, edited by Marina Calloni, Torino: UTET.

Anzoise, Valentina and Stefano Malatesta. 2010. "Visual and tourist dimensions of Trentino's borderscape." Pp. 44–61 in *Tourism and visual culture*, edited by Peter M. Burns, Jo-Anne Lester, and Lyn Bibbings. Wallingford, UK: CABI.

Beccali, Marco, Gussoni Maristella, and Francesca Tosi, eds. 2003. *Ergonomia e ambiente. Progettare per i cinque sensi. Metodi, strumenti e criteri d'intervento per la qualità sensoriale dei prodotti e dello spazio costruito*. Milano: Il Sole 24Ore-Pirola.

Bianchi, Elisa and Felice Perussia. 1978. *Il centro di Milano: percezione e realtà. Una ricerca geografica e psicologica*. Milano: Unicopli Editore.

Blaut, James M., George F. Mc Cleary, and América S. Blaut. 1970. "Environmental mapping in young children." *Environment and Behavior* 2:335–349.

Boatti, Antonello. 2007. *Urbanistica a Milano*. Novara: De Agostini Scuola Editore.

Bourdieu, Pierre. 1990. *In other words: Essays towards a reflexive sociology*. Cambridge: Polity Press.

Bourdieu, Pierre and Loïc Wacquant. 1992. *An invitation to reflexive sociology*. Cambridge: Polity Press.

Cavallini, Elena. 2005. "Lo spazio vissuto." Pp.132–142 in *Geografia a scuola: monti, fiumi, capitali o altro?*, edited by Marcella Schmidt di Friedberg. Milano: Guerini e Associati.

Chiesura, Anna. 2004. "The role of urban parks for the sustainable city." *Landscape and Urban Planning* 68:129–138.

Coley, Rebekah, Frances E. Kuo, and William C. Sullivan. 1997. "Where does community grow? The social context created by nature in urban public housing." *Environment and Behavior* 29:468–494.

dell'Agnese, Elena. 2005. *Geografia politica critica*, Milano: Guerini e Associati.

dell'Agnese, Elena and Valentina Anzoise. 2011. "Milan, the unthinking metropolis." *International Planning Studies* 16(3):217–235.

Downs, Roger M. and David Stea, eds. 1973. *Image and environment*. Chicago: Aldine Publishing Co.

Evans, Gary. 1980. "Environmental cognition." *Psychological Bullettin* 88:259–287.

Faccioli, Patrizia. 1997. *L'immagine sociologica. Relazioni famigliari e ricerca visuale*. Milano: Franco Angeli.

Faccioli, Patrizia and Giuseppe Losacco, eds. 2008. *Identità in movimento. Percorsi tra le dimensioni visuali della globalizzazione*. Milano: Franco Angeli.

Geipel, Robert and Marcello Cesa Bianchi, eds. 1980. *Ricerca geografica e percezione dell'ambiente*. Milano: Unicopli.

Gibson, James. 1979. *The ecological approach to visual perception*. Boston: Houghton Mifflin.

Gould, Peter. 1970. "On Mental Maps." Pp. 260–282 in *Man, space and environment: concept in contemporary human geography*, edited by Paul Ward English and Robert C. Mayfield. New York: Oxford University Press.

Grady, John. 1996. "The scope of visual sociology" in *Visual Sociology* 11(2):10–24.

Harper, Douglas. 2002. "Talking about pictures: a case for photo-elicitation." *Visual Studies* 17(1):13–26.

Helgren, David M. 1983. "Place name ignorance is national news." *Journal of Geography* 82:176–178.

Henny, Leonard M. 1986. "Theory and practice of visual sociology." *Current Sociology* 34(3):1–76.

Hurworth, Rosalind. 2003. "Photo-interviewing for research." *Social Research Update* 40:1–4.

Jay, Martin. 1988. "Scopic regimes of modernity." Pp. 2–27 in *Vision and visuality*, edited by Hal Foster. New York: The New Press.
Killworth, Peter D. and Bernard H. Russel. 1982. "A technique for comparing mental maps." *Social Networks* 3:307–312.
Kitchen, Robert M. 1994. "Cognitive maps: what are they and why study them?" *Journal of Environmental Psychology* 14(1):1–19.
Lowenfeld, Viktor and W. Lambert Brittain, eds. 1984. *Creatività e sviluppo mentale*. Firenze: Edizioni Giunti.
Lynch, Kevin. 1960. *The image of the city*. Cambridge, MA: MIT Press.
Malaguzzi, Loris. 1995. *I cento linguaggi dei bambini*. Bergamo: Edizioni Junior.
Martinotti, Guido. 1993. *Metropoli*. Bologna: Il Mulino.
Massey, Doreen B. 1994. *Space, place, and gender*. Minneapolis: University of Minnesota Press.
Massey, Doreen B. and Pat Jess, eds. 1995. *A place in the world? Places, cultures and globalization*. Oxford: Oxford University Press.
Mattioli, Francesco. 1986. "Gli indicatori visivi nella ricerca sociale: validità e attendibilità." *Sociologia e Ricerca sociale* 7(20):41–69.
Melucci, Alberto, ed. 1998. *Verso una sociologia riflessiva. Ricerca qualitativa e cultura*. Bologna: Il Mulino.
Murray, Henry A. 1943. *Thematic Apperception Test*. Cambridge, MA: Harvard University Press.
OECD. 2006. *Territorial review*. Milan: OECD Publishing.
Ricolfi, Luca. 1997. *La ricerca qualitativa*. Roma: Carocci.
Roskos-Ewoldsen, Beverly, Timothy P. McNamara, Amy L. Shelton, and Walter Carr. 1998. "Mental representation of large and small spatial layouts are oriented dependent." *Journal of Experimental Psychology: Learning, Memory and Cognition* 24(1):215–226.
Saarinen, Thomas F. 1966. *Perception of drought hazard on the Great Plains*. Research Paper No. 106. Chicago: Department of Geography, University of Chicago.
Saarinen, Thomas F. 1973. "Student view of the world." Pp. 148–161 in *Image and environment*, edited by R. M Downs and D. Stea. Chicago: Aldine.
Sanders, Ralph A. and Philip W. Porter. 1974. "Shape in revealed naps." *Annals of the Association of American Geographers* 64:254–267.
Sanesi, Giovanni, Raffaele Lafortezza, Mirilia Bonnes, and Giuseppe Carrus. 2006. "Comparison of two different approaches for assessing the psychological and social dimensions of green spaces." *Urban forestry & Urban Greening* 5:121–129.
Schroeder, Herbert W. 1991. "Preferences and meaning of arboretum landscapes: combining quantitative and qualitative data." *Journal of Environmental Psychology* 11:231–248.
Sholl, M. Jeanne. 1987. "Cognitive maps as orienting schemata." *Journal of Experimental Psychology: Learning, Memory and Cognition* 13(4):615–628.
Tuan, Yi-Fu. 1974. *Topophilia: a study of environmental perception, attitudes, and Values*. Englewood Cliffs, NJ: Prentice-Hall.
Tuan, Yi-Fu. 1975. "Images and mental naps." *Annals of the Association of American Geographers* 65:205–213.
Wagner, John, ed. 1979. *Images of information* Beverly Hills, CA: Sage.

9 Operations of Recognition
Seeing Urbanising Landscapes with the Feet

Christian von Wissel

> Caminante, no hay camino. Se hace camino al andar
> (Walker, there is no path. Paths are made by walking)
>
> Antonio Machado

INTRODUCTION

This chapter invites its readers for a walk in the metropolitan areas of New York and Mexico City as well as across the disciplines of art and social sciences. It does so with the aim to explore the materiality of the processes of urbanization and to discuss the practice of "seeing with the feet" as a visual-sensory research method for urban—here sub- and peri-urban—enquiry.

I will first revisit the artist Robert Smithson's *Tour of the Monuments of Passaic*, a creative-analytical "operation of recognition" (Marot 2006) in the (sub)urbanizing landscape of Passaic and Rutherford, in New Jersey, in 1967. During this excursion I will examine Smithson's practice of walking and "site-seeing" (Smithson 1966–67) as a poetic, that is 'world-making', method to access the "visual construction of the city" (Krieger 2004, 2006b). In other words, I will walk the path *made* by walking by which practitioners of urban space construct aesthetic relations with their socio-material urban environment. I will furthermore analyze the distinct perspective, described as "crystalline view-point" (Reynolds 2003:92), which Smithson's walking practice opened up on the suburban reality of his times.

Building upon Smithson's intervention, the second part of this chapter challenges the scope of operations of recognition by conducting a similar walk in the contemporary (peri-)urbanizing landscape of Tecámac, a municipality located in the Federal State of Mexico at the fringe of the metropolitan area of Mexico City. By converting the researcher her/himself into the subject of an experimental research situation, this second walk aims to show that operations of recognition provide a fruitful means for research to be responsive both to the "relational perspective" (Ingold 2000) of bodies navigating the urban environment and to the shifting texture and materiality of urban form and life.

MAKING SENSE OF URBANIZING LANDSCAPES

With more than half of the world's population now living in urban agglomerations, the process of global urbanization reconfigures not only our cities but also the planet and its societies as a whole (Lefebvre 2008; Davis 2009; UN habitat 2010). Anticipating the implications of such vast transformation of the foundations of human existence and dealing with them "on street level" of everyday life is becoming an essential aspect of the urban condition. As a result, living in today's—and tomorrow's—"Endless City" (Burdett and Sudjic 2007) is ever-more so a socio-spatial (urban) experience which authors like Rowe and Koetter (1978), in regard to urban planning, and de Certeau (1988), in regard to everyday (urban) life, have tried to describe with the notion of *bricolage*: as a continuously made and re-made assemblage of multiple ways of "working things out" and "finding ones way."

Such condition of a constant reworking of the "possibilities of urban becoming" (Simone 2004:3) challenges our ways of understanding contemporary urban life and form. Inhabitants, researchers and policy makers alike are increasingly confronted with dimensions of everyday life that lie "'in-between' the categories and designations" by which social sciences and politics tend to organize urban phenomena into dichotomies like legal/illegal, formal/informal, or movement/home (Simone 2004:2, 16). Way in into the "Urban Age," we still "see things incompletely" because of looking at them through conceptual frameworks that do not help to focus on the emerging reality of what Henri Lefebvre has prominently referred to as the "Urban Revolution" (2008[1970]:29). Hence, in order to shape our actions in response to the urbanizing world, it is of vital interest to sharpen our picture of the urban: it is by the way we *see* our cities that we develop the relevant thinking tools and operational capacities to actively intervene in designing the qualities of our urban and continuously urbanizing lives (Harvey 1996:38). Accordingly, it is essential not only to recognize the city "everywhere and in everything" (Amin and Thrift 2002:1) but also to make visible *how* we recognize such all-embracing urban condition. Through what eyes do we look at the process of worldwide urbanization, and what do we see?

Taking up this path toward enhancing our understanding of global "citification" (Häußermann et al. 2008:22), scholars are increasingly looking into the development of complementary research methods capable of "re-imagining the urban" (Amin and Thrift 2002) in a way that is responsive to the complex nuances of contemporary urban socio-spatial processes and formations. In this light, the following "walk" alongside Smithson's exploration of Passaic county and from there onward into the urbanizing landscape of the municipality of Tecámac aims at providing an opportunity to discuss—as well as to experiment with—the practice of walking as a possible method for *seeing* the urban. Regardless of the central role Smithson's work plays within land art, and despite its relevance for the study of visual culture/practice (cf. Reynolds 2003) and its inspiring impact on the theory of landscape architecture (cf. Careri 2001; Marot 2006), his walking practice has received little attention within the social sciences and the field of urban studies. Even more so, it has rarely been looked at from a methodological angle for doing research on the urban (sub- and/or peri-urban) condition. Setting out to participate in filling this gap, my objective is to highlight the potential and limitations of Smithson's ambulatory art practice which I would like to frame as a sensory-visual technique of enquiry that is particularly responsive to the texture and materiality of socio-spatial urban form.

MONUMENTS MADE BY WALKING AND SEEING (PASSAIC, 1967)

In September 1967, Smithson sets out to his *Tour of the Monuments of Passaic*, an exploratory excursion along New Jersey's river Passaic. During this walk, Smithson follows the construction site of a new highway, traverses an adjacent parking lot and visits a suburban shopping street.

Three months after the tour, in December 1967, the artist publishes a combined written and visual/graphic account of this suburban voyage in the journal *Artforum* (Smithson 1967b:68–74), which included an essay, six photographs, the reproduction of a reproduction of a painting and a "negative map" of the walk's location. In this account, Smithson provides a description of his particular art practice that consisted of visual and creative interventions for which Sébastien Marot (2006:70) has created the term "operations of recognition." These operations constitute a two-fold practice of *recognizing*—that is, analyzing—urban form consisting of (a) experiencing the environment's physicality through the act of walking and (b) re-creating the environment's meaning through poetic visual appropriation of its constituting elements. An extract of the beginning of Smithson's account reads as follows (Smithson 1967b:68–71):

> On Saturday, September 20, 1967, I went to the Port Authority Building on 41st Street and 8th Avenue. I bought a copy of the *New York Times* and a Signet paperback called *Earthworks* by Brian W. Aldiss. Next I . . . purchased a one-way ticket to Passaic . . . and boarded the number 30 bus of the Inter-City Transportation Co.
>
> I sat down and opened the Times . . . I looked at a blurry reproduction of Samuel F. B. Morse's Allegorical Landscape . . . the sky was a subtle newsprint grey . . . I read the blurbs and skimmed through Earthworks. The first sentence read, "The dead man drifted along in the breeze." . . . The sky over Rutherford was a clear cobalt blue, a perfect Indian summer day, but the sky in Earthworks was a "great black and brown shield on which moisture gleamed."

The bus passed over the first monument. I pulled the buzzer-cord and got off at the corner of Union Avenue and River Drive. The monument was a bridge over the Passaic River . . . Noon-day sunshine cinema-ized the site, turning the bridge and the river into an over-exposed *picture*. Photographing it with my Instamatic 400 was like photographing a photograph. The sun became a monstrous light-bulb that projected a detached series of 'stills' through my Instamatic into my eye. When I walked on the bridge, it was as though I was walking on an enormous photograph that was made of wood and steel, and underneath the river existed as an enormous movie film that showed nothing but a continuous blank.

From the banks of Passaic I watched the bridge rotate on a central axis in order to allow an inert rectangular shape to pass with its unknown cargo. The Passaic (West) end of the bridge rotated south, while the Rutherford (East) end of the bridge rotated north; such rotations suggested the limited movements of an outmoded world. "North" and "South" hung over the static river in a bi-polar manner. One could refer to this bridge as the "Monument of Dislocated Directions."

Along the Passaic River banks were many minor monuments such as concrete abutments that supported the shoulders of a new highway in the process of being built . . .

Nearby, on the river bank, was an artificial crater that contained a pale limpid pond of water, and from the side of the crater protruded six large pipes that gushed the water of the pond into the river. This constituted a monumental fountain that suggested six horizontal smokestacks that seemed to be flooding the river with liquid smoke . . . A psychoanalyst might say that the landscape displayed "homosexual tendencies", but I will not draw such a crass anthropomorphic conclusion. I will merely say, "It was there."

The texts and images by which the artist communicates his *Tour of the Monuments* present a thick, multilayered illustration of his experience and perception during the walk. The account weaves together the artist's bodily and intellectual intervention (his walking and writing) and interpretation (his seeing/reading) of the environment. It brings together both what he saw and what he sensed, as well as the sense he made of the physical situation by means of comparing the "perceived" with the "lived" urban (social) space he traversed (cf. Lefebvre 2009:40).

Pictured through the perspective of a first-person narrator, the text juxtaposes different layers of physical and imaginary realities. For example, Smithson repeatedly refers to the sky and light conditions of (a) the material scenery he is walking through, (b) the science-fiction novel *Earthworks* he read while on the bus and, (c) the reproduction of the painting *Allegorical Landscape* he saw in the newspaper. These layers are projected onto each other in the atmospheric reconstruction of the journey. Furthermore,

Smithson constantly relates his act of seeing to the landscape he sees by reflecting on the medium by which he does this visual intervention: his Instamatic 400 camera.

Smithson's primary intervention is elevating a series of elements he finds in the environment into "monuments." For example, through the act of visually appropriating himself of certain elements of the landscape, Smithson transforms the swing bridge crossing the river Passaic into the "Monument of Dislocated Directions," and a series of pipes into "a monumental fountain."

The argumentation line (as well as the tour's inherent timeline) by which Smithson walks—that is, reads and writes—through the suburban landscape runs from the initial affirmation of the mere existence of these monuments via the recognition of the particular landscape these monuments come to signify to the appreciation of the irreversible dissolution of this landscape into a state of "infinite disintegration" and "sullen dissolution" of its constituents (Smithson 1967b:74). All monuments are pictured as "ruins in reverse," that is as "all the new construction that would eventually be built" because in the suburban context buildings "don't *fall* into ruin *after* they are built but rather *rise* into ruin before they are built" (Smithson 1967b:72). Imagined by Smithson as "holes" that "define, without trying, the memory-traces of an abandoned set of futures," the monuments are presented as moments of absence yet full of meaning.

SITE-SEEING: THE (RE)MAKING OF LANDSCAPES

With his operation of recognition, Smithson renders into visibility what could easily be looked over. He intervenes in the environment by converting into landmarks what most passersby would regard as lacking any significance. By *making* the Monuments of Passaic through seeing (and naming) them, Smithson is able to (re)define the suburban sphere as the "negative map" that is the (back)ground from which these (positive) figures stand out in front (cf. Rowe and Koetter 1978: 64ff). This elevation of the unremarkable constituents of the environment into monuments is of great importance. It allows Smithson—and us, the readers of his text—to see the unremarkable as the remarkable expression of this very *unremarkableness*. By carving monuments out of the environment, the artist first captures the material context of the walk and then, consecutively, (re)constitutes it as a particular, here urbanizing, landscape.

Robert Smithson's creative intervention builds on how landscapes are formed out of our positioning in relation to them. Landscapes are made by the act of seeing. They are read into the environment by taking us out of it to some physical or imaginary vantage point from which we are able to look back onto them (Burckhardt 2006:33). Hence, addressing a material reality as a landscape implies two activities: (1) to situate oneself in relation to a given environment, and (2) to construct this relation through vision. Such act of visually *making* an environment out of positioning oneself in relation to it has been coined—in the context of vast urban agglomerations—the "visual construction of the megalopolis" (Krieger 2004). It describes how we continuously build and reconfigure within our imagination an archive of images that is produced directly by the urban environment or via visual representations of it (photographs, drawings, etc.). This mental archive represents the environment according to our understanding of it, thus constituting our (visual) "urban imaginary." As "reproductive and productive imagination" the continuously reworking of this urban imaginary becomes an essential capacity for society "to know and structure their everyday urban environment" (Krieger 2006b:111, translation by the author). The significance of such visual construction of the city ranges from the need for orientation to the production of meaning (ibid.:180).

When cross-referencing to each other the different skies present in the New Jersey scenery, the science-fiction novel *Earthworks*, and the newspaper reproduction of a painting, Smithson pictures how the language he elaborates to capture his experience is informed by bodily practice, contextual knowledge, and preexisting visual references. Between the lines, he describes how we construct the landscapes in which we live through multilayered perception: here in form of the artist's own sensory reception together with art historically informed and fictional references. His account shows how the visual construction of the urbanized landscape is at work with images constantly moving back and forth between the city image and the imagined city, that is, between the material urban space and the perceiver's virtual-visual urban space (cf. Krieger 2001, 2004, 2006a).

Smithson refers to this act of *landscape making* as the practice of "site-seeing" (Smithson 1966–67:340–345; 1967c:52–60). Site-seeing is an active, conscious mode of "direct perception" of "sense-data" (ibid.) by which objects and spaces of the everyday are transformed into sites (cf. Marot 2006:70; Careri 2002:156). This is done through the act of aesthetic appropriation, which consists of perceiving objects as meaningful representations of their surroundings. It is also done by renaming these elements and, thus, reconfiguring them and their contexts in new constellations of meaning (e.g., "The Monument of Dislocated Directions" for the swing bridge). This poetic practice of *site*-seeing is developed in direct opposition to the tourist's gaze of sightseeing, which recalls a merely passive contemplation of sights.

In a press release originally published some months before his tour to New Jersey, Smithson makes a significant annotation regarding the working of this landscape-making notion of site-seeing. The piece is called *Language to Be Looked At and/or Things to Be Read* and was initially written to support a drawing that showed a text in pyramidal shape. In the short note, Smithson (1967a:61, emphasis added) casually states that his "sense of language is that it is *matter* and not ideas." The materiality of language expressed in the note is a central theme in the artist's creative practice.

For Smithson, language is of physical matter that can be looked at, read through, and, consequently, walked about. In his New Jersey tour, he explores this extended notion of language: through physically experiencing the environment he reads the urbanizing landscape as a material language through which the socio-spatial situation is speaking to him.

This leads to two important conceptions regarding how to understand Smithson's operations of recognition: First, it points toward the fact that space is effectively read not only through "totalizing vision" (cf. de Certeau 1988) but also through "errant" movement (Careri 2002:20). This possibility of reading space is present in the very materiality of the walking act, that is, in the material knowledge the body is able to gain of the world through handling—here *feet*ling—it in practice (cf. Bolt 2010:30, drawing on Heidegger 1966). Analogue to other art practices that work directly with "the matter of thought," this ability to recognize the physicality *by walking through it* can be conceptualized as a practice of "material thinking" (Carter 2004).

In this light, walking is a process of sensory empirical knowing that is based on the phenomenological premise that perception implies more than just the visual sense. Following James Gibson's argument, Tim Ingold (2000:3) points out that perception "is not the achievement of a mind in a body, but of the organism as a whole in its environment . . . immanent in the network of all our sensory pathways that are set up by virtue of the perceiver's immersion in his or her environment." Thus, our perception of the environment comes about while being immersed in it, and, consequently, it has to be understood as a relational engagement with our lifeworld. This is what Ingold (2000:5, 153) describes as "dwelling perspective": a *practitioner's* point of view by which we make sense of the world through actually making the world out of bodily sensing it.

Building on this argument, we can state that landscapes are made not only by visually positioning us in relation *to* them, but also by bodily acting *within* them. "Landscapes," Ingold (2009: 333) suggests, "emerge as condensations or crystallizations of activity within a relational field." They are walked into life in relation to the human being walking her/himself into life and weaving her/his life into the landscape. The walker, thus, does not only have an experience of the physical surrounding but is constituted by and constituting this surrounding. Through the act of bodily exploratory movement the subject makes her/himself *and* her/his world. Addressing pedestrian locomotion in terms of such a practice that is both poetic (i.e., world-making) and aesthetic (i.e., world-perceiving)[1] allows us to recognize the relational activity that is "feeling one's way" "through a world that is itself in motion, continually coming into being through the combined action of human and non-human agencies" (Ingold 2000: 155).

The second important conception developed within Smithson's operations of recognition is the notion of the materiality of language. This understanding introduces the possibility to perceive the Passaic riverbank, highway, parking lot, and commercial street as a material text composed of a series of elements related to each other through a particular syntax. Once such a relation among the landscape's parts is established, Smithson's operations are capable of visually "cutting out" these elements in order to dislocate and relocate them and, by doing so, to read and rewrite the landscape. By doing so, his practice follows the poetic and playful possibilities inherent in the collage as one of visual art's crucial modes of creation. In this light, Smithson's operations of recognition are as imaginative as they are creative in perceiving differently. The site-seeing artist does not try to order the complexity of the images produced by the suburban landscape into one rational, linear reading but embraces their tension by placing them in an allegorical play of mirrors and reflections (Reynolds 2003:82, 116). Within this imaginary assemblage, Smithson's operations are capable of establishing visual, spatial and, most importantly, temporal relations between the landscape's distinct elements (the different monuments and types of monuments) and, in his words, their different "stages of futurity" (Smithson 1967b:72; cf. Reynolds 2003:102).

THE CRYSTAL LAND: CONCEPT FORMATION IN THE URBAN SPHERE

Having identified the practice of site-seeing as the visual-sensory tool of data collection and analysis by which Smithson walks into being the meaningful relations he establishes with the environment, it is of similar interest to have a closer look at how Smithson pictures the suburban reality out of his relational perspective. Referring once again to David Harvey's claim (1996:38) regarding the importance of identifying how we see the urban world in order to design our interventions within it, the question is what alternative vistas of the processes of urbanization are opened up by Smithson's operations of recognition?

Most importantly, in this context, is the fact that suburbia is not captured by a single view opening up before the perceiver's eyes in the manner of a Renaissance or Romantic landscape painting. No clearly defined perspective and vanishing point help the act of recognition. On the contrary, Smithson (1967b:72, 70) describes the landscape as "zero panorama" and tries to capture it as a series of "cinema-ized," "over-exposed" still images of a moving picture. His reading of the landscape is that of a collage of sites: an assemblage of parallel existing realities but with different stages of futurity which in turn become manifest in the monuments. In other words, Smithson sees the environment not primarily as a landscape of spatial relations but as one constituted by traces and correlations in time.

From there on he reads the landscape as "absence," as a mirror and its reflection, both infinitely signifying one another (Smithson 1967b:74). The suburban landscape is pictured as a "text *without* subject, a representation without reference" (Marot 2006:80, emphasis added). It is mere appearance, without the physical source from which this game of eternal referencing starts. In Smithson's (1967b:72) words, the urbanizing landscape is full of "holes," places where past, presence and future merge into different stages of timeless futurity: the ruin in reverse that rises from the future back to the present, free of history yet full of memory of the process of becoming. It is a "utopia minus a bottom," composed of the physical manifestations of "monumental vacancies." This leads him to perceive Passaic as a land of low density and unremarkableness, as an "unimaginative" and "clumsy eternity" when compared to the "tightly packed and solid," "interesting" New York City (Smithson 1967b:72–73).

However, when Smithson describes the landscape as *being there without being there* he directs the view to the processuality of urbanization. Suburbia is characterized by showing the remains of an abandoned future—that is, by materializing what is left over of what has not yet been—capturing in its built appearance the constant possibility of a "city yet to come" (cf. Simone 2004). In Smithson's account of Passaic, he shows suburban form as a perpetually unfinished presence, as a state of tinkered (*bricolent*) improvisation and temporality: "always in the making," in endless repetition of itself. The urbanizing landscape is seen as the "eternal" state of an irreversible process leading to infinite dissolution (Smithson 1967b:74); eternal in the sense that urbanization never comes to a hold in producing a somehow finished entity called "the city" but continuously reproduces the sub and peri-urban condition of temporal, spatial and conceptual "in-betweenness." Passaic, understood in the light of this ever-repeating, self-mirroring becoming as a kind of "New Rome," a new 'eternal city', is represented in what Smithson site-sees as the monument to a "model desert": a sand box he finds along the way to the end of his tour.

Besides these readings, there are additional connotations hidden between the lines of the text. Eye catching, for example, is the fact that both the description of the situation and the analytical tools for its interpretation are almost entirely centered on spaces and concepts of movement. It is a "moving picture" Smithson's Instamatic 400 projects into his eyes, and the experiences he describes are either made on the bus or walking along the (unfinished) highway, the parking lot, and the shopping street of Passaic city—this last one he describes as a "no centre," a "typical abyss or an ordinary void" because of its transitory character and which could need a place-making "'outdoor sculpture show' [to] pep that place up" (Smithson 1967b:70).

Furthermore, Smithson recognizes the landscape as being self-made by its suburban inhabitants. The ruins in reverse are all but one the products of infrastructure projects conceived to guarantee the mobility needs of modern, car-based, everyday (urban) life. Making an ironic comment on his own—and ours, the readers'—involvement in the phenomenon of vast suburbanization, Smithson finds on a sign beside the unfinished road a remark which says that it is "*your* Highway Taxes" that are at work in producing this "particular kind of heliotypy (Nabokov), a kind of self-destroying postcard world of failed immortality and oppressive grandeur" (Smithson 1967b:72, emphasis added; cf. Reynolds 2003:112).

In analytical terms, Smithson's contributions to the understanding of urbanizing landscapes lie in a series of perceptual shifts, which allow him to render remarkable the unremarkable through materially thinking the environment—that is, by perceiving it by means of seeing it "with the feet." Furthermore, with his operations of recognition he was one of the first to explore and make visible the immanent potentials of the new protagonists of urbanizing form as they massively reproduced themselves in the American sprawl of the 1960s: the formless, undefined spaces of suburbia which were experienced as possessing neither city-like nor rural qualities (cf. Careri 2002:180).

The academic debate of his time regarding the "exploding metropolis" (Reynolds 2003:84)—a debate in which Smithson participated with a lecture on *Shaping the Environment: The Artist and the City* in 1966—perceived suburbanization as a problem of "imageability," that is, as the lack of visual references needed by the inhabitants in order to construct notions of belonging (cf. Lynch 1960). Following Aldo Rossi's argument, the suburban landscape was perceived as missing the existence of *historic* monuments as these where understood to be the necessary visual anchors that could materialize "persistence" and thus identity (cf. Rossi 1982[1966]).

In this context, Smithson's first contribution was literally filling the gap by (simply) creating the called-for monuments. However, the way he approached the solution was very different to the "historicist and picturesque contextualism" (Krieger 2006b:240) emerging in postmodern architecture and urbanism in response to Rossi's and Lynch's claims. Contrary to searching for—or reinventing—historic references of an arguably permanent state of persistence, Smithson defined some of the landscape's elements as *future* monuments that serve as the reference points of a process of urban becoming. In other words, rather than "getting lost" amid the lack of history as he was supposed to according to the theorists of his time (Reynolds 2003:84), he turned the supposed deficit into a creative opportunity: site-seeing the memory traces of the future of diffuse suburban space. Doing so, Smithson established positive vantage points from which to look at the situation in order to read it while at the same time maintaining the relational perspective of being placed within and being part of the environment. Instead of introducing monuments that are "frontal expressions" of dominant relations of social production (cf. Lefebvre 2009:33), the land artist walks and sees counter-hegemonic monuments into existence as analytical elsewheres (cf. Reynolds 2003) from which to (re)access the situation in order to understand it.

Smithson's second contribution was the shift away from perceiving the urbanizing landscape from a single viewpoint perspective. In order to understand the—then—new phenomenon of urbanization and increased mobility, theorists and artists favored the notion of the car view—looking out through the windshield of a moving car—as the relevant perspective by which modern society was establishing its relation with the environment (Reynolds 2003:89; cf. Appleyard and Lynch 1964; Smith 1966; Venturi et al. 1968). Here, Smithson confronts the dominant reading with an alternative. He refers to the New Jersey landscape as "The Crystal Land" (Smithson 1966:7), reading the suburban reality as a reflection/representation of itself and thus developing what Ann Reynolds (2003:92) calls the "crystalline view-point." This viewpoint is not moving toward a fixed vanishing point, even though Smithson, too, elaborates his operation of recognition within the framework of the car and highway as the two phenomena that transformed the experience of urban space and time. Instead, Smithson substitutes the lineal perspective with the decomposing and recomposing possibilities of the collage. In opposition to the single vanishing point, which he describes as being "pointless" (Reynolds 2003:95), the crystalline viewpoint presents the advantage to be composed of multiple foci structured by "punctuation" (Smithson 1967a:61). Following the land artist, this way of seeing is highly sensitive to the unfolding suburban sphere because of its capacity to create the landscape out of three interconnected modes of aesthetic appropriation: (1) by recognizing the surrounding from within multiple viewpoints and in relation to multiple vanishing points, (2) by anchoring the elements registered out of these diverse perspectives as the markers (monuments) of the particular socio-material and temporal context, and (3) by relating these elements/monuments to each other as if they were the sentences and subordinate clauses of a visual text. Applying and combining these three operational modes, the practitioner of suburban space makes sense of the urbanizing environment precisely through reading its spatial and temporal multiplicities, and through writing these multiplicities into crystalline—instead of lineal-singular—relations.

The third contribution and, likewise, important perceptual shift lies in Smithson's careful reading, and his spirited appraisal, of the complexity of modern everyday life. Smithson not only captured the contradictions and multiplicities of the suburban environment but also embraced the emerging reality's immanent uncertainties as "paradoxical qualities of everyday modern life" (Reynolds 2003:8). By means of "collapsing the experience of the future, present, and past together," Smithson frees his descriptions from falling into the "corrupted memory"—as he puts it—of an imagined "natural" landscape which others have lamented about as some kind of "lost state of purity" previous to the dissolution of "the city" into the entropic process of urbanization (Reynolds 2003:117–118). In other words, in his understanding of the New Jersey landscape, Smithson accepts the profound transformations of the human condition caused by the process of urbanization. His way of seeing allows anticipating the shifting experiences of an unfolding Urban Age.

SEEING URBANIZING LANDSCAPES WITH THE FEET (TECÁMAC, 2010)

As outlined above, Smithson's walking and site-seeing practice shows to be a mobile, sensory-visual research intervention that allows "walking into vision" the materiality of the continuously urbanizing world. Smithson's practice of "seeing with the feet" is a technique of carefully sensing the texture of what surrounds us. By carving monuments out of the unremarkable, his poetic-aesthetic walking practice allows recognizing the material (back) ground of urban everyday life.

Smithson's *Tour of the Monuments of Passaic* further suggests the possibility to experience in practice the relational perception of the environment by which practitioners in urbanizing space develop an understanding of the urbanizing landscape in which they operate. In order to scrutinize these potentials, an experimental walk was conducted in the municipality of Tecámac, in the northern continuum of the metropolitan area of Mexico City. Building on the method and theoretical concepts developed by Robert Smithson in the 1960s, the artist-researcher's operations of recognition where adopted as a tool for visual-sensory data collection and analysis responsive both to the aesthetic, visual-sensory construction of the city and to the constant becoming of urban, socially produced, material space.

URBANIZATION IN THE METROPOLITAN AREA OF MEXICO CITY

At the beginning of the twenty-first century, the ongoing process of sub- and peri-urbanization is a global phenomenon that in many aspects resembles the situation walked about by Robert Smithson in the 1960s. Contemporary accounts of urban agglomerations, and of the experiences to have within these agglomerations, recall the artist's description of suburbia as an *urbanizing future eternally repeating (reflecting) itself into being* (cf. Smithson 1967b:74; cf. Soja 2000; Marot 2006). Following Walter Prigge (1998), one can certainly ask if, rather than the city, by now it is the periphery that is everywhere? In this light, Smithson's concepts describing urbanizing form can be read as a prophesy of Lefebvre's (2008) notion of the Urban Revolution, materializing itself in vast, entropic sub and peri-urban landscapes that reinvent the urban as much as they change the face of our planet (cf. Soja 1992:95; cf. Davis 2009).

The metropolitan area of Mexico City is without doubt a vast urbanizing landscape that contains all the attributes of the emerging, all-embracing—in Smithson's words "eternal"—city of the Urban Age. It is an agglomeration that houses more than 20 million people, living in 76 to 104 boroughs and municipalities spreading out across three federal states (SEDESOL et al. 2007; COESPO 2009—the different numbers come about through competing definitions and a continuously expanding geographical

172 *Christian von Wissel*

scope regarding the metropolitan area). Consequently, the agglomeration has long ago left to be a traditionally comprehensive city unit. In fact, the city has never existed as one single entity but emerged out of an urban system of pre-Hispanic and colonial cities, towns, and villages which, over the past 100 years, has further developed into a complex metropolis and regional megalopolis (cf. Ward 1991; Eibenschutz 1997; Garza 2000; Garza and Schteingart 2010). Inhabitants of the surrounding two federal states outnumber the inhabitants of the region's inner federal state—the Distrito Federal—by eleven to nine million (CONAPO 1998).

At the northern fringe of this urbanizing system, formal and informal growth has created a peri-urban continuum where rural, suburban, industrial, and urban forms co-exist in constant tension and/or juxtaposition (cf. Aguilar and Ward 2003; Nivón 2005). In the midst of this becoming *bricolage* of urban form and life we find the municipality of Tecámac and, at its heart, the Carretera México-Pachuca highway where the second walk presented in this chapter takes place.

WALKING AND SITE-SEEING THE PERI-URBAN CONTINUUM

As stated before, the objective of my research intervention is to challenge Smithson's operations of recognition in regard to their viability as a method for visual-sensory urban research. Furthermore, by employing Smithson's walking method, I seek to access the urban practitioner's—here, the researcher's—walking perspective on the urbanizing landscape. In order to do so, I take my own practice of ambulatory and poetic seeing as both the tool and the subject of my enquiry. In other words, I set out for an exploratory *Tour of the Monuments of Tecámac* by slipping into Smithson's shoes of the "emplaced" (urban) researcher, being "her- or himself part of a social, sensory and material environment" (Pink 2009:23)—an environment, which I aim to walk into vision. Following Smithson's path, and paraphrasing his writing, my experimental operation of recognition begins:

> On Saturday, January 02, 2010, I went to the Indios Verdes Bus Terminal at the northern end of Insurgentes Avenue. I found my way to platform B and boarded the next bus to Tecámac where I purchased a ticket from the driver . . . The sky over Ecatepec was cobalt blue with patches of glaring grey; a typical winter day in the Valley of Mexico.
>
> The bus passed the first monument. I approached the driver and asked him to let me off outside the housing estate Rancho La Luz. The monument was the pylon of an over-land electricity line crossing the street and adjacent estate and farmland . . . Photographing the environment through its metal beams with my D-Lux 4 was like slicing it into triangle-shaped fragments. When I walked around the pylon, it was as though I was walking around an enormous Aleph[2] that contained all

Operations of Recognition 173

the images of the patchwork urbanising landscape of the peri-urban continuum of the Metropolitan Area of Mexico City . . .

Along the Carretera México-Pachuca highway were many minor monuments such as concrete manholes that equipped the sewage ductwork of the road widening in the process of being built . . . It was hard to distinguish the broadening from the road; they were both confounded into a unitary chaos.

Nearby, on the street's edge, was a pile of worn out tyres . . . This constituted a monumental ziggurat marking a reference point of human activity along the way . . . A psycho-geographer might say that the landscape revealed its moving tendencies yet I will merely say, "It was there."

Off the bus, I start walking and seeing, that is, "seeing with the feet." I do so, along what James Gibson (1986) has called the "path of observation": recognizing the urbanizing landscape out of my own relational, exploratory perception (cf. Ingold 2004:331). My "circumambulatory knowing" of the world allows me to site-see the unremarkable traces of human urban activity as remarkable monuments of the socio-spatial and constantly materializing processes of urbanization.

The first such landmark I come across is what I call "The Pylon Monument" (Figure 9.1). The picture my camera projects into my eye resembles how the Mexican Valley is urbanized in the manner of an endless accumulation of disconnected fragments.

One housing estate next to the other slices the former agricultural farmland into pockets of enclosed, discrete suburban realities: a shattered landscape, each shard resembling the other yet neatly cut apart by walls and (electric/barbed) wires. In Smithson's terms, it is a crystalline land repeating itself in the endless dispersal of both its particles and its energy; with

Figure 9.1 The Pylon and the Bank Account Monuments.

each mirror piece being connected to the wider urban context only by being threaded onto the string of the highway that functions as the regions one and only "power supply."

Further on, I find several minor monuments right beside the street in the unpaved, "soft" hard shoulder of the road. I give the first one the name "The Manhole Monument" as it is an entry shaft to a new sewage duct that has been brought into the ground yet lacks completion (Figure 9.2).

Rising above the ground by ten centimeters, the monument invites me—and the reader—to see two important aspects regarding how urbanizing landscapes are made out of the spatial practitioner's ambulatory perception: on the one hand, my operation of recognition renders visible the processuality of urbanization by drawing attention to the physical manifestation of the urbanizing landscape's condition of being "in-the-making." Guiding me along Smithson's concepts, I address the unfinished entry shaft as the "memory-trace" of an "abandoned set of futures" (Smithson 1967b:72). Doing so, the Manhole Monument invites the user of this particular space to envision urban modernization as an utopian project, that is, as a landscape of urban becoming rising above the ground of the unfinished street in the form of an imaginary paved street to be completed at the level of the upper rim and cover of the shaft. At the same time, seeing the manhole as a "monument" while walking along it allows me to make explicit the mode by which specific constituents of the landscape take on significance. It is by literally falling over it with my eyes and feet that the Manhole Monument provides the opportunity to visually and sensory recognize how space is qualified out of the perspective of living it (cf. Ingold 2000:153). In other words, the urbanization process of Tecámac becomes apparent in the practitioner bodily experiencing the material condition of such citification and visually appropriating her/himself of it.

Figure 9.2 The Manhole Monument and Monumental Line.

Operations of Recognition 175

Furthermore, the monument provides a reference for visual orientation within the environment. Drawing on Krieger (2006b:180), "urban monuments"—either conventionally defined or site-seen in the manner of Smithson's creative appropriation—"still function as effective models for the production of 'meaning' in standardised urban landscapes." As individually and collectively made constituents of the "urban image," they originate and determine how the city is visually constructed in the urban practitioner's imaginary (Krieger 2006b:26). A research activity consisting of seeing monuments, therefore, allows the social scientist to access such "visual construction of the city" (Krieger 2004). Yet, while the heightened level of attention involved in such purposeful site-seeing is assumed to provide a detailed recognition of the visual impact of the environment's materiality, this is not to say that this materiality does not also play an important role in the everyday formation of meaningful sensory, bodily relations between inhabitants and their lived-in surrounding. Rather, the visual construction of the city operates beneath "those forms of recall that the actor is able to express verbally" (Giddens 1986:48), that is, at the level of the consciousness of the body and of the possibility of material knowing (cf. Rhys-Taylor 2010:11–12; Carter 2004). It is within this light, that Peter Krieger states that it is a daily socio-cultural exigency for every inhabitant, "conscious or not," to "relate to each other and make sense out of these multiple and heterogeneous elements that integrate urban space" (Krieger 2006b:231, translations by the author).

In addition to the visual-sensory relation, the monuments along the Carretera México-Pachuca highway also reveal spatial-temporal relationships between the urban practitioner and the material landscape. "Down here," in the "present tense" of the sewage duct's new entry shaft pocking out of the ground, the Manhole Monument speaks of the surface texture and

Figure 9.3 The Tyre and Pyramid Monuments.

material condition of what it means to move—that is, to walk or drive—in this particular urbanizing situation. It does so together with a set of other minor monuments I find to the left and right of the road. These are "the Tyre Monument (also called The Ziggurat)," "the Monumental Line"—a line made of stones traversing the unpaved shoulder of the street—and "the Pyramid Monument"—a pile of three stones left also in the middle of the highway's side-strip (Figures 9.2 and 9.3).

Following Tim Ingold (2000:193) in his argument regarding the formation of landscapes from *within*, these monuments help to describe the urbanizing landscape as "the world as it is known to those who dwell therein, who inhabit its places and journey along the paths connecting them." They give testimony of the material experiences that make up living in the urbanizing landscape of the Central Mexican Valley, that is, of the condition of the urban practitioner's physical emplacement by which the landscape is both registered *on* and "grown" *through* the body (Ingold 2000:85). These experiences range from the dust that stirs up around me while walking on the unpaved surface, to the traces of human (nomadic and sedentary) activity left behind along the way: dumped flat tires, neatly drawn divisions of space and deliberately made small-scale landmarks (or is the Pyramid Monument the memory-trace of an improvised lifting jack?). Again, the scope of the experiment is not to ascribe ultimate meaning to any of these constituents of the landscape, but to recognize their impact on the ambulatory construction of meaningful relations between acting individual and enacted space.

Ingold (2004:333) suggests that "the forms of the landscape—like the identities and capacities of its human inhabitants—are not imposed upon a material substrate but rather emerge as condensations or crystallizations of activity within a relational field." The monuments of Tecámac, thus, are more than pure symbolic transformations by which the urban practitioner—here the researcher—visually constructs the landscape (cf. Krieger 2006b:8). They are also the marks of "human beings living *in* the world, not *on* it" and, as such, they speak of the ongoing transformations effected by this "being in the world" as "part and parcel of the world's transformation of itself" (Ingold 2004:333, emphasis added). The last monument I site-see during my walk further supports this approach to making sense of Tecámac's peri-urban environment. It suggests a creative strategy for saving which is why I name it "the Bank Account Monument (also known as "the Monument of Impro-(in)vestment")" (Figure 9.1). It consists of a wall made of loosely piled-up cement stones that serve as both enclosure and physical savings account: unused building material (the leftovers of a concluded or the remains of an abandoned construction activity?) is stored for future usage, yet, at the same time, it is put to work in the present as the totally functional physical delimitation of a car wreck merchant's property. This financial-reserve-equals-wall monument allows recognizing the urbanizing landscape as the work of *bricolage*, as the material expression of a "hands-on" practice of improvisation, that is, of "working-things-out" (cf.

de Certeau 1988). By site-seeing the wall as a bank account, the process of urbanization becomes visible as the urban practitioner's material ability of making the (urban) world through *handling* it in practice (cf. Bolt 2010).

OPERATIONS OF RECOGNITION: A VISUAL-SENSORY RESEARCH TOOL FOR URBAN ENQUIRY

As the visit to Smithson's walk in New Jersey and the short exemplary excursion in Tecámac tried to demonstrate, this chapter argues that operations of recognition present a responsive method for understanding sub and peri-urban landscapes out of the bodily and visual relation one establishes with their materiality.

Framed within the endeavor to re-imagine the urban, revisiting Robert Smithson's *Tour of the Monuments of Passaic* shows fertile, I argue, due to the accomplished (re)positioning of the (artist)/researcher's body in the research field, as well as due to the relational perception such (re)positioning enables. Smithson centers his method of analysis in the urban practitioner's mobile and poetic immersion within the environment. By doing so, his operation of recognition allows him to reveal the *bricolant* (making-do) conditions of the urbanizing landscape from *within* the very transformation of the environment. When site-seeing the construction of the new highway on the Passaic riverbank into a series of monuments, Smithson walks into recognition the process of constant urban becoming, that is, he comes to know urbanization through the bodily practice of bipedal locomotion and relational perception. Thus, by sensing the environment through walking, Smithson is able to experience—and, therefore, to materially make sense of—the transformation of the peri-urban continuum from within its materiality.

This is confirmed by the second walk in the Mexico City's peri-urban continuum: the monuments I walked into being with my *Tour of the Monuments of Tecámac* all allowed me to experience and picture the processuality of the metropolitan landscape of the Valley of Mexico. Site-seeing monuments provided access to the poetic and interpretative skills that urban practitioners need, and have, at hand and foot to recognize, make sense of and—out of this relational practice—"grow" the landscape they inhabit. Above all, it allowed sensing and visualizing the role played by the environment's physical texture in the process of an active, sensory-visual construction of the city, that is, by way of seeing, materially thinking, and operating in (everyday) urban space.

Such grounded and embodied perspective on the urbanizing world constitutes a method of "seeing with the feet" that allows the researcher to recognize how practitioners in urban space come to relate to the physical conditions of the environment through modes of both visual and bodily perception. Putting into action Robert Smithson's (1967c:56) claim for using "the actual land as a medium" for commenting on the world, the artist discovers the materiality of the territory as a source for formulating relevant concepts of how to relate to and, thus, to make sense of the environment. Smithson immersed with all his senses into the realm of the suburban reality. By doing so, he was among the first to respond with his body to the process of (accelerating) urbanization.

Examining the technique by which I collected and analyzed my data in Tecámac, I would like to further discuss the potential of walking and site-seeing as a combined research method for social urban enquiry. I argue that operations of recognition show their prospective contribution to methodologies in the visual urban sphere in a number of ways.

First, operations of recognition provided a useful framework for reconfiguring the viewpoint by which to access the phenomenon of urbanization. The notion of a collage-like and cinematized "crystalline viewpoint" helps drawing the attention toward the possibility of reading the urbanized landscape as a "punctuated" whole in which certain elements—the monuments or ruins in reverse—articulate the meaning of the urbanizing environment. Robert Smithson's excursion into the materiality of language shows to be a fruitful tool in order to read the texture of a given urban situation. Just like a text is not simply a lineal flow of words but acquires meaning through structuring these words into a composition of propositions, (urbanizing) landscapes are not to be seen as one fixed view toward a lineal vanishing point but as a set of highly complex "crystallizations of activity" (Ingold 2004:333), woven into meaningful relations through the act of walking. Smithson's approach of what we could call a "linguistics of the material language" helped to make visible the actively composed nature of landscapes as well as their materiality's impact on the urban image production which Peter Krieger (2004; 2006a:350) refers to as the aesthetic construction of the metropolis, rooted in visual-sensory perception.

Second, operations of recognition help to focus on the practice of walking in itself. In the context of shifting and often contradictory physical outcomes of the process of urbanization, the experiment in Tecámac proved walking to be a useful practice to find access to this complexity. As Tim Ingold and Jo Lee Vergunst (2008) have pointed out, walking is not a lineal process but a mode of wayfaring and thus way-finding. Walking, therefore, shows to be responsive to the equally fluid conditions of *bricolage*, 'in-betweenness," improvisation and constant "in-the-making" that the urbanization process brings into being in form of hybrid rural-industrial-suburban-urban—in one word: urbanizing—landscapes.

Third, operations of recognition allow thinking research as a sensory experience in which the 'emplaced' researcher (Pink 2009:23) actively immerses in the texture of the studied situation. Gaining material knowledge of the urban sphere by being "in touch" with the physicality of its everyday life "through our feet" (Ingold 2004:330) helps to access what Tim Ingold (2000:5) describes as "dwelling perspective," that is, our perspective of ecological relations when we understand ourselves as being bodily embedded and actively engaged with the constituents of our surrounding. Handling—here *feetling*—the world in practice (Bolt 2010:29) allows to touch the physicality, the socio-material "textureness," of the urban experience within the relational perspective by which we interact with the urbanizing landscape we live in. Building on this, employing operations of recognition as a research method provides a first step toward restoring touch, smell, hearing, and taste, as well as the senses of balance, thermo-sensitivity and proprioception—the sense we have of our own body in space—to their proper place in social sciences research of the urban sphere (cf. Ingold 2004:330; Rhys-Taylor 2010:10–11, drawing on Simmel 1997:110). Here, a path is laid out toward further research into the sensory-spatial dimensions of everyday (urban) life.

The principle limitation of operations of recognition becomes apparent when Michael Bull and Les Back (2003:1–2) remind us of how the visual "has often meant that the experience of the other senses . . . has been filtered through a visualist framework." Smithson sees the landscape as photographs and moving images. His operations of recognition are clearly subject to the primacy of vision. In Smithson's artistic research practice, seeing—although accompanied by the other senses—is the filtering access point through which we are invited to immerge into the environment and to physically sense its texture when walking about it.

WALKING ON

Operations of recognition provide a responsive method for seeing and handling the environment in order to explore the role that the physical surrounding is playing in the formation of meaningful relations between social practitioner, practice, and socially produced material space. Their contribution to a "Sociology of the Visual Sphere" lies in drawing our attention to the *matter* of space, that is, to both the materiality of the (urban) world and to the visual-sensory construction by which actors in space attach/retrieve meaning from this materiality. Drawing on anthropology, sociology, and art practice as research, operations of recognition further the notion of a mobile and multi-sensory poetic perception by which space is "made" for and through the living body (cf. Lefebvre 2009:169).

Building upon these capabilities, Smithson's technique of walking perception provides the opportunity to set out further research on how urban practitioners construct particular local (sub- and/or peri-urban) notions of belonging when trying to make sense with the senses of the material constituents of their lifeworlds. Focusing on the construction/growth of relationships with the material aspects of the environment, operations of recognition promise to unfold their full potential when employed alongside complementary research tools that highlight the social production of space. As a method to access the process of a visual-sensory making of the world from within personal experience, operations of recognition have to be extended toward research practices that allow evaluating other (urban) practitioners' relational perceptions.

Smithson's land art practice described here deliberately aims to experience the relationship between the sub-/peri-urban practitioner and her/his landscape practiced within. It employs a combination of walking and seeing as an operational tool for researching everyday urban space from within the "research operator's" dwelling perspective. In contrast to de Certeau's (1988:93) notion of the walking body writing an urban "text" "without being able to read it," Smithson's operation of recognition write *and* read urban space by experiencing and site-seeing its materiality (cf. Careri 2002:156). Seeing urbanizing landscapes with the feet, thus, constitutes a practice of mobile visual-material recognition of (social) space allowing us to think the urban from within its physical and visual sphere.

NOTES

1. When referring to the aesthetic dimension of the practice of walking my intention, following on Krieger (2006b, 350), is to recollect "the original meaning of the word aisthesis in Ancient Greek, which includes all the facets of sensorial perception and the intellectual understanding thereof.
2. In his short story, The Aleph, Jorge Luis Borges describes the aleph as "one of the points in space containing all points", as a "place where there are, without confusion, all places of the world, viewed from all angles." (Borges 1997, 187-8, translation by the author). He explicitly draws on kabalistic meanings attributed to the Aleph being the first letter in the Hebrew alphabet ('Aleph bet'). He also refers to set theory, where the Aleph "is the symbol for transfinite numbers, for which the whole is not greater than any of its parts" (ibid, 196).

REFERENCES

Aguilar, Adrian Guillermo, and P. M. Ward. 2003. "Globalization, regional development, and mega-city expansion in Latin America: analyzing Mexico City's peri-urban hinterland." *Cities* 20(1):3–21.

Amin, Ash and Nigel Thrift. 2002. *Cities—reimagining the urban*. London: Polity.

Appleyard, Donald, Kevin Lynch, and John R. Myer. 1964. *The view from the road*. Cambridge, MA: MIT Press.

Bolt, Barbara. 2010. "The magic is in handling." Pp. 27–33 in *Practice as research: approaches to creative arts enquiry*, edited by Estelle Barrett and Barbara Bolt. London: I. B. Tauris.

Borges, Jorge Luis. 1997[1949]. "El Aleph." Pp. 175–198 in *El Aleph*. Madrid: Alianza. English: Borges, Jorge Luis. 2004. *The Aleph and other Stories*. London: Penguin.

Bull, Michael, and Les Back, eds. 2003. *The auditory culture reader*. Oxford: Berg.

Burckhardt, Lucius. 2006. *Warum ist Landschaft schön? Die Spaziergangswissenschaft*, edited by Markus Ritter and Martin Schmitz. Berlin: Schmitz.

Burdett, Richard, and Deyan Sudjic, eds. 2007. *The endless city*. London: Phaidon.

Careri, Francesco. 2001. *Walkscapes: El Andar Como Prática Estética*. Barcelona: Gustavo Gili.

Carter, Paul. 2004. *Material thinking: the theory and practice of creative research*. Carlton, Victoria: Melbourne University Press.

Certeau, Michel de. 1988[1980]. *The practice of everyday life*. Berkeley: University of California Press.
COESPO—Consejo Estatal de Poblacion del Estado de Mexico, ed. 2009. *Zona Metropolitana del Valle de Mexico*. Secretaria General de Gobierno del Estado de Mexico. Retrieved July 2, 2011 (http://www.edomex.gob.mx/poblacion/docs/2009/PDF/ZMVM.pdf).
CONAPO—Consejo Nacional de Población (Mexico). 1998. *Escenarios demográficos y urbanos de la Zona Metropolitana de la Ciudad de México, 1990-2010: síntesis*. México: Consejo Nacional de Población.
Davis, Mike. 2009. "Who will build the ark? The architectural imagination in an age of catastrophic convergence." Pp. 23–35 in *Culture nature: art and philosophy in the context of urban development*, edited by Anke Haarmann and Harald Lemke. Berlin: Jovis.
Dell, Christopher. 2007. "Die Performanz des Raumes." *archplus: Situativer Urbanismus* 183:136–143.
Eibenschutz Hartman, Roberto. 1997. *Bases para la planeación del desarrollo urbano de la ciudad de México: Estructura de la ciudad y su región*. México D.F.: UNAM u.a.
Garza, Gustavo, ed. 2000. *La Ciudad de México en el fin del segundo milenio*. México: El Colegio de México, Centro de Estudios Demográficos y de Desarrollo Urbano; Gobierno del Distrito Federal.
Garza, Gustavo, and Martha Schteingart, eds. 2010. *Desarrollo urbano y regional*. México D.F.: Colegio de México.
Gibson, James. 1986[1979]. *The ecological approach to visual perception*. Hillsdale, NJ: Lawrence Erlbaum.
Giddens, Anthony. 1986. *The constitution of society: outline of the theory of structuration*. Berkeley: University of California Press.
Harvey, David. 1996. "Cities or urbanization?" *City* 1:38–61.
Häußermann, Hartmut, Dieter Läpple, and Walter Siebel. 2008. *Stadtpolitik*. Bonn: Bundeszentrale für Politische Bildung.
Ingold, Tim. 2000. *The perception of the environment: essays on livelihood, dwelling and skill*. London: Routledge.
Ingold, Tim. 2004. "Culture on the ground: the world perceived through the feet." *Journal of Material Culture* 9(3):315–340.
Ingold, Tim, and Jo Lee Vergunst, eds. 2008. *Ways of walking: ethnography and practice on foot*. Aldershot, UK: Ashgate.
Krieger, Peter. 2001. "Desamores a la ciudad satellites y enclaves." Pp. 587–606 in *Amor y desamor en las artes. (XXIII Coloquio Internacional de Historia del Arte)*, edited by Arnulfo Herrera. Mexico, DF: Instituto de Investigaciones Estéticas, UNAM.
Krieger, Peter. 2004. "Construcción visual de la megalópolis México." Pp. 111–139 in *Hacia otra historia del arte en México: Disolvencias (1960-2000) Tomo IV*, edited by Issa María Benítez Dueñas. Mexico, DF: Consejo Nacional para la Cultura y las Artes.
Krieger, Peter. 2006a. "Citambulation: Distinguish, understand and exploit the imaginaries of the Megacity of Mexico." Pp. 346–359 in *Citamblers: guide to the marvels of Mexico City; the incidence of the remarkable*, edited by Ana Álvarez, Valentina Rojas Loa, and Christian von Wissel. Mexico, DF: Conaculta and Cultura sin Fondos.
Krieger, Peter. 2006b. *Paisajes Urbanos: Imagen Y Memoria*. Mexico, DF: Universidad Nacional Autónoma de México; Instituto de Investigaciones Estéticas.
Krieger, Peter. 2009. "Aesthetics and anthropology of megacities: a new field of art historical research." In *Histoire de l'art et anthropologie*, Paris, edited by INHA and Musée du quai Branly (« Les actes »), [Online]. Accessed March 27, 2010 (http://actesbranly.revues.org/318).

Lefebvre, Henri. 2008[1970]. *The urban revolution*. Minneapolis: University of Minnesota Press.
Lefebvre, Henri. 2009[1974]. *The production of space*. Oxford: Blackwell Publishers.
Lynch, Kevin. 1960. *The image of the city*. 29th ed. Cambridge, MA: MIT Press.
Marot, Sébastien. 2006[2003]. *Suburbanismo y el arte de la memoria*. Barcelona: G. Gili.
Nivón Bolán, Eduardo. 2005. "Hacia una antropología de las periferias urbanas." Pp. 140–167 in *La antropología urbana en México*, edited by Néstor García Canclini. México D.F.: Consejo Nacional para la Cultura y las Artes; Universidad Autónoma Metropolitana; Fondo de Cultura Económica.
Pink, Sarah. 2009. *Doing sensory ethnography*. London: Sage.
Prigge, Walter, and Stiftung Bauhaus Dessau, ed. 1998. *Peripherie ist Überall*. Edition Bauhaus. Frankfurt/Main: Campus.
Reynolds, Ann. 2003. *Robert Smithson: learning from New Jersey and elsewhere*. Cambridge, MA: MIT Press.
Rhys-Taylor, Alex. 2010. "Coming to our senses: a multi-sensory ethnography of class and multiculture in East London." Doctoral thesis. Goldsmiths, University of London. Retrieved April 2, 2011 (eprints.gold.ac.uk/3226/1/Alex_Rhys_Taylor.pdf).
Rossi, Aldo. 1982[1966]. *The architecture of the city*. Cambridge, MA: MIT Press.
Rowe, Colin, and Fred Koetter. 1978. *Collage city*. Cambridge, MA: MIT Press.
SEDESOL, CONAPO, and INEGI. 2007. *Delimitación de las zonas metropolitanas de México 2005*. México D. F.: Secretaría de Desarrollo Social; Consejo Nacional de Población; Instituto Nacional de Estadística, Geografía e Informática.
Simmel, Georg. 1997. *Simmel on culture: selected writings*, edited by David Frisby and Mike Featherstone. London: Sage.
Simone, AbdouMaliq. 2004. *For the city yet to come: changing African life in four cities*. Durham, NC: Duke University Press.
Smith, Tony. 1966. "Talking with Tony Smith (interview by Samuel Jr. Wagstaff)." *Artforum*, December:18–19.
Smithson, Robert. 1966. "The crystal land." *Harper's Bazaar*, May. Reprint. Pp. 7–9 in *Robert Smithson, the collected writings*, edited by Jack Flam. 1996. Berkeley: University of California Press.
Smithson, Robert. 1966–1967. "The artist as site-seer; or, a dintorphic essay." Pp. 340–345 in *Robert Smithson, the collected writings*, edited by Jack Flam. 1996. Berkeley: University of California Press.
Smithson, Robert. 1967a. "Language to be looked at and/or Things to be read." Press release. New York: Dwan Gallery. Reprint. P. 61 in *Robert Smithson, the collected writings*, edited by Jack Flam. 1996. Berkeley: University of California Press.
Smithson, Robert. 1967b. "The Monuments of Passaic." *Artforum*, December. Reprint. Pp. 68–74 in *Robert Smithson, the collected writings*, edited by Jack Flam. 1996. Berkeley: University of California Press.
Smithson, Robert. 1967c. "Towards the development of an air terminal site." *Artforum*, June. Reprint. Pp. 52–60 in *Robert Smithson, the collected writings*, edited by Jack Flam. 1996. Berkeley: University of California Press.
Soja, Edward. 1992. "Inside exopolis: scenes from Orange County." Pp. 94–122 in *Variations on a theme park*, edited by M. Sorkin. New York: Noonday Press.
Soja, Edward. 2000. *Postmetropolis: critical studies of cities and regions*. Oxford: Blackwell Publishers.
United Nations Human Settlements Programme. 2010. *State of the world's cities 2010–2011: bridging the urban divide*. London: Earthscan.
Venturi, Robert, and Denise Scott Brown. 1968. "A significance for A&P parking lot, or learning from Las Vegas." *Architectural Forum*, March. Pp. 37–43.
Ward, P. 1991. *México: una megaciudad. Producción y reproducción de un medio ambiente urbano*. México, DF: Alianza; Consejo Nacional para Cultura y las Artes.

Contributors

Valentina Anzoise has a PhD in Information Society. Since 2003 she has collaborated with the Visual Research Lab at Milano-Bicocca University. She is author of several publications on visual and cultural studies, environmental perception and representation, innovation and sustainability policies, and the development of interdisciplinary approaches and participatory techniques. Currently she is a Research Associate at the European Centre for Living Technology at Ca' Foscari University of Venice (Italy). She is involved in the Emergence by Design research project, funded by the European Commission under the FP7-ICT-FET programme

Jerome Krase is emeritus and Murray Koppelman Professor, Brooklyn College, of the City University of New York. He has published and lectured extensively on urban communities, and has been photographing urban neighborhoods around the world for more than thirty years. His most recent book is *Seeing Cities Change: Local Culture and Class* (Ashgate 2012) and he co-edits *Urbanities*, the journal of the Commission on Urban Anthropology < http://www.anthrojournal-urbanities.com/>

Ayelet Kohn is currently working at Department of Photographic Communication Hadassah Academic College, Jerusalem, Israel. Her main research interests are multimodality and its uses in social contexts. She has published in journals *Visual Communication, Multicultural Education, Emergencies: Journal for the Study of Media and Composite Cultures, Journal of Tourism and Cultural Change*, and more. E-mail: ayeletkohn@gmail.com

Cristiano Mutti has a PhD in Information Society. Since 2000 he has coordinated activities of the Visual Research Lab at Milano-Bicocca University. He is author of several publications on visual and qualitative techniques, and his main research interests are visual and cultural studies, and the integration of different media, tools, and techniques for the development of interdisciplinary and qualitative research approaches.

184   *Contributors*

Regev Nathansohn is the co-founder (with Dennis Zuev) of the Visual Sociology Thematic Group working under the International Sociological Association (ISA), and serves as its president (2010–2014). He has published articles on the visual sociology of the Israeli Occupation comparing practices of professional photojournalists to those of amateur photographers. Regev is currently a PhD candidate in the department of Anthropology at the University of Michigan (Ann Arbor), and teaches visual anthropology at Hadassah Academic College in Jerusalem. His current research focuses on concepts and practices of "coexistence" in a "mixed" (Arab-Jewish) neighborhood in Israel, and on the politics of their visual representations. Email: regev@umich.edu

Łukasz Rogowski is assistant professor at the Institute of Sociology, Adam Mickiewicz University in Poznań. His sphere of research interests encompasses visual sociology (both as a visual methodology and sociology of visual culture), sociology of the Internet, social theory, and researching socio-cultural competencies. His doctoral dissertation ("Social Visual Competence as a Research Problem of Contemporary Sociology") won the Polish Prime Minister Award for Outstanding Doctoral Dissertations in 2011. E-mail: lukasz.rogowski@gmail.com

Anna Schober received her PhD (in 2001) as well as her postdoctoral habilitation (in 2009) in contemporary history at the University of Vienna. She is currently Mercator Visiting Professor at the Justus Liebig University Giessen (financed by the DFG. Deutsche Forschungsgemeinschaft) and was guest researcher at several international institutions such as Verona University, Italy; University of Essex, Colchester, UK; the J.W. Goethe University in Frankfurt, Germany; the IFK (International Research Center for Cultural Studies) in Vienna; and the Jan van Eyck Academy in Maastricht, Netherlands. She has published widely on the aesthetic and political dimensions of the public sphere, popular culture, new social movements, and gender studies.

Timothy Shortell is associate professor of sociology and director of the MA program in sociology at Brooklyn College, of the City University of New York. He has published numerous articles and chapters on social semiotics of the public sphere. He is currently working on a visual study of immigrant neighborhoods in Brooklyn and Paris.

Pavithra Tantrigoda is a lecturer (probationary) at the Department of English, University of Colombo, Sri Lanka. She is currently reading for her PhD in literary and cultural studies at Carnegie Mellon University, Pittsburgh. Her research interests include transnational fiction, law and literature, critical theory, and the theories on the visual.

Matteo Vergani has a PhD in sociology and the methodology of social research. His current interests focus on the use of digital technologies by political activists. He currently is a teaching assistant at the Catholic University of Milan.

Christian von Wissel is an urbanist and architect living in Mexico City and Berlin. He is co-author of *Citámbulos: The Incidence of the Remarkable, Guide to the Marvels of Mexico City* (Oceano, 2007) and co-curator of the exhibition *Citámbulos: Through the Looking Glass, Journey to the Mexican Megalopolis*. Currently, he is doing his PhD in visual sociology at the Centre for Urban and Community Research at Goldsmiths, University of London.

Dennis Zuev graduated from the Krasnoyarsk State University, Russia, and received his PhD in sociology of culture from Altay State University, Russia, in 2004. Currently he is a research fellow at the Centre for Research and Studies in Sociology, CIES-ISCTE, IUL in Lisbon, Portugal. He is involved in two research projects: Conditions and Limitations of Lifestyle Plurality in Siberia, funded by the Max Planck Institute for Social Anthropology, Germany; and "Selfing": Contact, Magic and the Constitution of Personhood, funded by Fundação para a Ciência e a Tecnologia (FCT), Portugal. E-mail: tungus66@gmail.com

Index

A

activism, 57, 75, 99
ambivalence, 66, 75–76
Anders, Władysław, 47
Anderson, Elijah, 125
androgyny, 5, 62–63, 64, 66, 68, 74, 77 n17
audiovisual data, 131

B

Baer, Ulrich, 16, 19, 22
Baranovich, Nimrod, 84, 99
Barat, Haji, 95
Barbie, 72–74
Barney, Matthew, 70
Barthes, Roland, 36, 66
Baudrillard, Jean, 13, 15, 20, 53
Beauregard, Robert A., 111
Beauty of Loulan, 95
Berlin, 47, 52, 73–74, 116, 118, 123, 125
body, 35, 38, 59, 61, 63, 66–72, 74–76, 77nn16–17, 78n21, 140, 166, 175–176, 177–180;
  body parts, 66, 76;
  hermaphrodite, 63, 66
Borer, Michael, 126
Bourdieu, Pierre, 42, 111, 131
Brooklyn, 109, 116, 117, 118, 119, 120, 123, 125
Burgin, Victor, 42

C

Campbell, David, 14, 19
Cape Town, 109, 116, 119, 120, 122
Castoriadis, Cornelius, 61, 62
certainty, 59–60, 75
de Certeau, Michel, 49, 74, 161, 166, 177, 180

Chase, John, 48
China, 83–84, 88, 92, 94, 97, 99–100, 102–103, 104n2, 105n8, 105n15, 105n20–21, 106n28, 106n30
  anti-Chinese, 96, 99, 105n6
Christianity, 31, 45–46
clown, 74–76
CNN, 19, 21–22
communication, 38, 42, 43, 45, 50, 57, 83–85, 91, 92, 96, 100–101, 103–104, 113–115, 117, 126, 126n2, 129
communism, 47
conflict, 3, 4, 6, 13–23, 23n1–3, 31–39, 47, 62, 64, 66, 69, 75, 76, 83, 86, 90, 94, 98, 104n2, 119, 134–135, 137
control, 15, 17, 38, 44–45, 70, 76, 108, 112, 126
Constantine the Great, 45
consumer culture, 72
conversion, 60–61
Comani, Daniela, 65–66, 70
Crary, Jonathan, 42
Crawford, Margaret, 48

D

Derderian, Richard L., 121
DeSena, Judith N., 125
destruction, 4, 15, 18, 21, 23, 43–50, 52, 54n7
Diaspora, 17, 83, 86, 88, 90, 91, 99, 101, 103–104, 104n2, 106n30
distrust, 4, 43–44
Drozdowski, Rafał, 52
Dzerzhinsky, Felix, 47–48, 54n5

E

Edwards, Elizabeth, 42

Emmerich, Roland, 49
Emmison, Michael, 53n2
Eroticism, 68, 74, 76, 77n18
ethnicity, 83, 84, 93, 96, 97, 100, 104, 104n2, 108–111, 117–121, 124–125, 126n2, 134
 ethnic identity, 84, 113, 117, 119, 121
ethnography, 116, 126,
European Union, 58
everyday life, 2, 46, 52, 60, 95, 108–109, 111, 114, 124, 137, 161, 168, 170, 171, 178–179

F
Feldman, Allen, 14–15
feminism, 57–59, 69, 74, 77n5
 feminist movement, 63
 femininity, 5, 35, 60, 62, 70, 72, 74, 77n16
folksonomy, 86
Frankfurt, 116, 120
Freedberg, David, 43, 46
Fritzsche, Peter, 111
Frosh, Paul, 37
fundamentalism, 85, 96, 104

G
Garland-Thomson, Rosemarie, 42
Gehl, Jan, 122
gender, 2, 4–5, 43, 57–76, 76n1, 77n9, 77nn11–13, 77n18, 87, 88, 124, 126n2, 157n36
 gender mainstreaming, 57, 58
 sex-gender, 57, 59, 62, 75, 76n4
 transgender, 59, 62, 68, 69
globalization, 6, 108, 109, 112, 118, 123, 125
 Alterglobalization movement, 49, 74–75
glocalization, 6, 109, 118, 124–125
Goffman, Erving, 42, 115
Gothenburg, 109, 116, 117, 123, 125, 155n1
Gottdiener, Mark, 113
Griffin, Michael, 13, 16

H
habitus, 111
Haila, Anne, 111
Hall, Stuart, 13, 14, 21, 43, 90
Harper, Douglas, 116, 129, 131
Hart, Janice, 40
Harvey, David, 112, 161, 167
Hawkes, Terence, 114

Hirsch, Marianne, 42
Hoxha, Enver, 47, 54n6
Hum, Tarry, 111
humor, 3, 25–39, 88
Hussein, Saddam, 54n7
hypersexual, 68, 69, 77n16, 77n18

I
iconic figuration, 62–63
iconoclasm, 4, 5, 42–53, 53n4, 54n7, 54n16
ideological codes, 6, 8, 90–104
identity, 6, 7, 8, 42, 57, 59, 62, 70, 71, 77n17, 84, 99, 103, 108–126, 126n1, 134, 169
image, power of, 2–5, 13–23, 44, 59–66, 72, 75–76, 130, 134, 137
imagery complex, 97–98
immigrants, 28, 110, 118, 121–124, 135, 156n29
Internet, 6, 17, 27, 46, 49, 50, 53, 69, 83, 85, 101, 103, 104
interpretive communities, 90–91
irony, 27, 29–30, 34–35, 38
Islam, 53n4, 84, 86, 91, 95, 96, 98, 101, 104, 110, 120

J
Jackson, John B. 111
Jakobson, Roman, 114
Jay, Martin, 26, 37, 50, 130
Al Jazeera, 19, 20, 21
Jesus, Christ, 45

K
Kadeer, Rabbiya, 86, 91, 96, 99
Kaliski, John, 48
Karner, Tracy X., 115
Katriel, Tamar, 36
King, Anthony D., 111
Kirshenblatt-Gimblett, Barbara, 28
Kirshner, Micha, 26, 33–34, 39n2
Krase, Jerome, 110, 111, 115, 119

L
Lalvani, Suren, 42
Lange, Patricia, 9n3, 85
Latour, Bruno, 43, 44
Lefevbre, Henri, 111, 161, 163, 169, 171, 179
Lenin, Vladimir Ilyich, 47
Levac, Alex, 26, 32, 39n2
Liberation Tigers of Tamil Eelam (LTTE), 3, 13–20, 23nn1–3

Lisbon, 116, 122, 123
Lobnor, 92
Lofland, Lyn H., 111, 112, 113, 125, 126
Los Angeles, 109, 111, 116, 119–120
Lynch, Kevin, 111, 130, 137, 169, 170

M

Mahsum, Hasan, 86, 95, 98, 102
Manchester, 109, 116, 117, 120, 121–122, 123
Marin, Louis, 61
masculinity, 5, 60, 62, 65, 72, 75, 77n16
materiality, 2, 4, 42, 160, 162, 165, 166, 171, 175, 177–178, 179–180
media, 2, 3, 13–23, 23nn1–2, 25, 38, 43, 46, 47, 49, 52, 59, 83–84, 87–89, 91, 102, 103, 113, 118, 131
mental maps, 129–131, 134, 138–141, 144–145, 147, 148–149, 153–154, 156n19
Messaris, Paul, 43
Milan, 8, 45, 116, 123, 132, 133, 135–137, 141–146, 148–149, 152, 153, 155nn6–7
Militant, 16–17, 91, 92, 96–104
Mirzoeff, Nicholas, 18
Mitchell, William J.T., 34, 42, 44, 52, 61
monument, 8, 43, 47, 52, 53n2, 119, 160, 162–178
Moses, 44, 45
Multiculturalism, 109, 119
myth, 17, 60–62, 66, 68, 77n17, 91, 96, 105n7

N

narrative, 13–15. 16, 17, 19, 20, 21–23, 26, 27, 30, 32, 36, 60, 102, 104
Nathansohn, Regev, 31, 35, 37
nationalism, 5, 83, 84, 90, 92, 94, 96, 101
network, 6, 49, 83, 85, 89, 166
neutral, 18, 20, 51, 66, 76
neutrality, 63–64
New Britain, 116, 121
New York, 50, 52, 116, 121, 160, 162, 168
norm, 5, 27, 37, 39, 42, 61, 62, 69, 72–74, 76, 77n9, 85

Nowotko, Marceli, 47

O

Olaf, Erwin, 75
operation of recognition, 160,164,170,172,174,177,180
Oslo, 116, 118
Osterhammel, Juergen, 109

P

pain, 22,76
Pan, Lin, 128
Paris, 51, 109, 116, 118, 119, 120, 121, 123
parity, 59, 64, 66, 74
participation, 85, 135
Passaic (New York), 160, 161, 162, 163, 164, 166, 168, 171, 177
passion, 59
patchwork self, 69, 70, 76
Pauwels, Luc, 9n3, 115
perception, 129, 130, 132, 133, 137, 141, 145, 149, 153, 159, 163, 165, 166, 177, 178, 179, 181
St. Petersburg, 116, 119, 125
Petersson, Niels P., 109
Philadelphia, 109, 116, 117
photo elicitation, 7, 129, 131, 141, 144, 146, 151, 153
photography, 16, 18, 19, 23–27, 29, 30, 38, 39, 40, 53, 55, 65
photographs, 3, 5–7, 13–17, 19, 22, 24–27, 29, 30, 37–39, 48, 53, 69, 92, 95, 102, 109, 115, 116, 141, 148, 153, 162, 165, 179
photojournalism, 25,26
photographer's gaze, 3,26, 36, 38, 39
picture, 6, 13, 18, 42–46, 48–53, 53n2, n4, 54n7, 63, 70, 97, 100, 102, 134, 137, 141, 143, 144, 146, 149, 152, 156 n32, 161, 163–165, 167, 168, 169, 173, 177
pictogram, 63, 64, 80
politics, 1–3, 19, 20, 22, 27, 58, 84, 88, 98, 100, 107, 125, 161
political criticism, 25, 36,
political grassroots movement, 57, 59
privatizing, 66
public life, 58, 61, 66, 77n11

Q

Qualitative Research, 90

190 Index

Quantitative Research, 87, 103
queer, 58, 59

R
Reformation, 46
relational perspective, 160, 167, 169, 179
religion, 31, 42, 45, 53 n4, 97, 108
representation, 2, 3, 7, 13, 14, 16, 17, 19, 20, 21, 25, 30, 31, 38–40, 43, 44, 46, 55, 59, 61, 62, 65, 72, 74, 77, 84, 85, 90, 91, 96, 103, 111, 113, 114, 129–131, 135, 136, 138, 146, 153, 154, 155n17, 165, 167, 170
Ritzer, George, 125
Robertson, Roland, 109
Rose, Gillian, 42

S
Sautman, Barry, 84
Saville, Jenny, 66–69, 78n19
scopic regimes, 14–15, 37
semiotics, 1, 6, 109, 110, 113, 114, 116, 126
senses, 4, 9, 50–51, 53, 54n15, 109, 178–179
sexual difference , 59, 62
sexuality, 5, 59, 62, 71, 75
Sherman, Cindy, 69, 70, 75
Shortell, Timothy, 111, 115
sight, 1, 4, 8, 20, 35, 42, 43, 45, 46, 50–53, 53n2, 112, 113, 126, 130, 165
signs,6,7, 28, 108, 109, 11, 113–116, 118–126, 126n2, 126n4, 152
    conative signs, 114, 123, 125
    expressive signs, 114, 115, 119–121, 124, 125, 125n2
    phatic signs,114, 115, 116, 118, 119, 123,124
    poetic signs, 114, 121–123, 125, 125n2
Simmel, Georg, 111,113,179
smell, 51, 113, 179
Smithson, Robert, 8, 160–162, 163–170, 171–174, 175, 177–180
social agency,114
social construction, 58
social movement, 125
social reform, 57
solidarity, 29, 63, 96, 98, 101, 104, 122
Sontag , Susan, 14, 16, 18, 22
Sperber, Dan, 26

Sri Lanka, 3, 13, 15–18, 20, 21, 23n1, 23n3
Stalin, Joseph, 47
Stockholm, 116, 121
Strauss, Anselm, 113
streetscapes, 8, 111, 115
Süskind, Patrick, 50
Świerczewski, Karol, 47
symbolic interactionism, 6, 109, 112, 116
symmetry, 64, 66, 74, 77n13
    asymmetry, 74, 76
system transformation, 46

T
Tarim mummies, 94, 105n14
Tecámac (Mexico City), 8, 160, 161, 171, 172, 174, 176–178
trauma, 22, 23
Turkestan, 86, 92, 95, 96–98, 104n3, 104n8, 105nn19–21
    Pan-Turkism, 84, 93, 107
    pan-turkist, 84, 92–96, 99
Tusk, Donald, 48
Tyner, Kathleen, 43

U
United Nations Meeting on the Status of Women in Beijing 1995, 58
Urban 8, 48, 49, 108- 116, 121–126, 126n4, 129, 130, 132, 137, 145, 146, 152–153, 155 n3, 156 n31, 160–163, 165, 167–180
    urban landscape, 2, 6, 129, 175
    peri-urban, 146, 148, 160, 162, 168, 171–173, 176, 177, 179–180
    suburbia, 167–169, 171
    urbanising landscape, 8, 160, 161, 166, 168–169, 171–174, 176–177, 179–180
    urban life, 8, 48, 108, 110, 113, 114, 126, 161, 177–180
    urban materiality, 160, 162, 171, 179
Uyghur, 5, 6, 83–104, 104nn2–3, 105n8, nn15–16, nn20–21, n25, 106n28
Uyghurstan, 86, 104n3

V
vernacular landscape, 111, 113–116, 123–127
vídeo, 5–6, 9, 13–22, 23n2, 48, 49, 54 n7, 57, 77n16, 83–85, 86–91,

92–97, 98, 99, 100, 101–102, 103, 104, 105n8, nn11–12, n21, n24
Visual
  visual construction of the city, 160, 175, 177
  visual data, 3, 6–7, 13, 18, 20, 101, 109, 116, 125, 126, 145
  visual markers, 69, 91
  visual production, 68–69, 101
  visual-sensory research, 160, 171–172, 177
  visual sociology, 1, 9, 9n1, 42, 54n15
  visuality, 1, 2, 42–46, 50, 52–53, 53n2
  the limits of, 13, 18, 20

W

Walking, 1, 124, 160–163, 166, 168, 171–174, 176, 178, 179, 180, 181
War, 13–15, 16, 17, 18, 19–23, 23nn1–3, 33, 47, 54n7, 64, 95, 96
Warren, Carol A.B., 115
Web 2.0, 5, 85

Weibel, Peter, 55
witness, witnessing, 13, 14, 20, 21, 23, 23n1, 38, 39, 69, 102
Wittgenstein, Ludwig, 59
Wohl, Richard R., 113
Wolberg, Pavel, 26, 35, 37, 39n2

X

Xinjiang, 83–84, 86, 88, 92, 96, 97, 100, 104n3, 105n14, 106n30

Y

YouTube, 2, 5, 6, 54 n10, 83–92, 96, 101–104, 104n5, 105nn8–13, nn15–24, nn26–27, 106n29, nn31–32

Z

Zelizer, Barbie, 31
Zerilli, Linda, 59, 62
Žižek, Slavoj, 21
Zukin, Sharon, 111